The NASA STI Program Office ... in Profile

Since its founding, NASA has been dedicated to the advancement of aeronautics and space science. The NASA Scientific and Technical Information (STI) Program Office plays a key part in helping NASA maintain this important role.

The NASA STI Program Office is operated by Langley Research Center, the lead center for NASA's scientific and technical information. The NASA STI Program Office provides access to the NASA STI Database, the largest collection of aeronautical and space science STI in the world. The Program Office is also NASA's institutional mechanism for disseminating the results of its research and development activities. These results are published by NASA in the NASA STI Report Series, which includes the following report types:

- TECHNICAL PUBLICATION. Reports of completed research or a major significant phase of research that present the results of NASA programs and include extensive data or theoretical analysis. Includes compilations of significant scientific and technical data and information deemed to be of continuing reference value. NASA counterpart of peer-reviewed formal professional papers, but having less stringent limitations on manuscript length and extent of graphic presentations.

- TECHNICAL MEMORANDUM. Scientific and technical findings that are preliminary or of specialized interest, e.g., quick release reports, working papers, and bibliographies that contain minimal annotation. Does not contain extensive analysis.

- CONTRACTOR REPORT. Scientific and technical findings by NASA-sponsored contractors and grantees.

- CONFERENCE PUBLICATION. Collected papers from scientific and technical conferences, symposia, seminars, or other meetings sponsored or co-sponsored by NASA.

- SPECIAL PUBLICATION. Scientific, technical, or historical information from NASA programs, projects, and missions, often concerned with subjects having substantial public interest.

- TECHNICAL TRANSLATION. English-language translations of foreign scientific and technical material pertinent to NASA's mission.

Specialized services that complement the STI Program Office's diverse offerings include creating custom thesauri, building customized databases, organizing and publishing research results ... even providing videos.

For more information about the NASA STI Program Office, see the following:

- Access the NASA STI Program Home Page at *http://www.sti.nasa.gov*

- E-mail your question via the Internet to help@sti.nasa.gov

- Fax your question to the NASA STI Help Desk at (301) 621-0134

- Phone the NASA STI Help Desk at (301) 621-0390

- Write to:
  NASA STI Help Desk
  NASA Center for AeroSpace Information
  7121 Standard Drive
  Hanover, MD 21076-1320

NASA/TM-2006-214273

Aerocapture Systems Analysis for a Titan Mission

*Mary K. Lockwood, Eric M. Queen, David W. Way, Richard W. Powell, Karl Edquist, Brett W. Starr,
Brian R. Hollis, E. Vincent Zoby, and Glenn A. Hrinda*
NASA Langley Research Center, Hampton Virginia

Robert W. Bailey, Jeffery L. Hall, Thomas R. Spilker, Muriel Noca, and Robert Haw
Jet Propulsion Laboratory, Pasadena, California

Carl G. Justus and Aleta Duvall
Computer Sciences Corporation, Huntsville, Alabama

Dale L. Johnson
NASA Marshall Space Flight Center, Marshall Space Flight Center, Alabama

James Masciarelli
NASA Johnson Space Center, Houston, Texas

Naruhisa Takashima
AMA Inc., Hampton, Virginia

Kenneth Sutton
National Institute of Aerospace, Hampton, Virginia

Joe Olejniczak and Bernard Laub
NASA Ames Research Center, Moffett Field, California

Michael J. Wright and Dinesh Prabhu
ELORET Corporation, Sunnyvale, California

R. Eric Dyke
Swales Aerospace, Hampton, Virginia

National Aeronautics and
Space Administration

Langley Research Center
Hampton, Virginia 23681-2199

February 2006

The use of trademarks or names of manufacturers in the report is for accurate reporting and does not constitute an official endorsement, either expressed or implied, of such products or manufacturers by the National Aeronautics and Space Administration.

Available from:

NASA Center for AeroSpace Information (CASI)
7121 Standard Drive
Hanover, MD 21076-1320
(301) 621-0390

National Technical Information Service (NTIS)
5285 Port Royal Road
Springfield, VA 22161-2171
(703) 605-6000

TABLE OF CONTENTS

Titan Aerocapture Systems Analysis	1
Titan Aerocapture Mission and Spacecraft Design Overview	13
Titan Explorer Mission Trades from the Perspective of Aerocapture	24
Approach Navigation for a Titan Aerocapture Orbiter	35
Engineering-Level Model Atmospheres for Titan and Neptune	47
Guidance Algorithms for Aerocapture at Titan	53
Aerocapture Simulation and Performance for the Titan Explorer Mission	61
Preliminary Aerothermodynamics of Titan Aerocapture Aeroshell	72
An Analysis of the Radiative Heating Environment for Aerocapture at Titan	83
Thermal Protection Concepts and Issues for Aerocapture at Titan	92
Structural Design of the Titan Aerocapture Mission	102
Planetary Probe Mass Estimation Tool Development and its Application to Titan Aerocapture	110

TITAN AEROCAPTURE SYSTEMS ANALYSIS

Mary Kae Lockwood

NASA Langley Research Center

INTRODUCTION

Performance projections for aerocapture show a vehicle mass savings of between 40 and 80%, dependent on destination, for an aerocapture vehicle compared to an all-propulsive chemical vehicle. In addition aerocapture is applicable to multiple planetary exploration destinations of interest to the NASA Office of Space Science. These results led to the identification of aerocapture as one of the top three propulsion technologies for solar system exploration missions during the 2001 NASA In-Space Propulsion Program (ISP) technology prioritization effort, led by Marshall Space Flight Center, to rank current ISP propulsion technologies. An additional finding was that aerocapture needed a better system definition and that supporting technology gaps needed to be identified.

An aerocapture systems analysis effort was kicked off in late February and completed in September 2002. The focus of the effort was on aerocapture at Titan with a rigid aeroshell system. Titan was selected as the initial destination for the study due to potential interest in a follow-on mission to Cassini/Huygens. The systems analysis is being completed by a multi-center NASA team including scientists and engineers from Ames Research Center, the Jet Propulsion Laboratory, Johnson Space Center, Langley Research Center, and Marshall Space Flight Center, led by Langley Research Center. Continued aerocapture systems analysis work is in progress with a Neptune aerocapture systems analysis study. Neptune is representative of the gas giant planets. Additional destinations and further work will be defined based on NASA Office of Space Science roadmap updates and ISP technology development. Plans in FY04 include Mars and Venus.

SYMBOLS/NOMENCLATURE

A	Area (m^2)
AFE	Aeroassist Flight Experiment
α_{trim}	Trim Angle of Attack
BOC	Beginning of Cassini
CA	Axial Force Coefficient
CBE	Current Best Estimate
CD	Coefficient of Drag
CFD	Computational Fluid Dynamics
CG, cg	Center of Gravity
CL	Coefficient of Lift
CN	Normal Force Coefficient
D	Drag
EGA	Earth Gravity Assist
EOC	End of Cassini
HGA	High Gain Antennae
L	Lift
L/D	Lift-to-Drag ratio
M/CDA	Ballistic Coefficient (kg/m^2)
SEP	Solar Electric Propulsion
TPS	Thermal Protection System
VGA	Venus Gravity Assist

BACKGROUND

An aerocapture flight profile schematic is shown in Figure 1. The vehicle approaches the planet/moon from a hyperbolic approach trajectory, shown at point 1, designed to achieve state conditions including

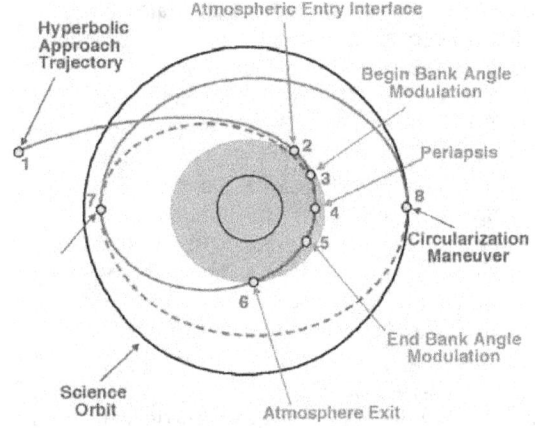

Figure 1. Aerocapture flight schematic.

flight path angle at atmospheric interface, point 2, within a predetermined range. Bank angle modulation, rotation of the lift vector about the velocity vector is initiated by the guidance at point 3. The drag on the vehicle as it passes through the atmosphere provides the delta V required to capture the vehicle into the desired orbit. The amount of delta V imparted to the vehicle is controlled by the onboard guidance by modulating bank angle, i.e. the direction of the vehicle's lift vector. A command of lift up during an atmospheric pass results in increasing altitudes, nominally decreasing atmospheric density, reduced drag and reduced delta V imparted. A command of lift down results in decreasing altitudes, nominally increasing atmospheric density, increased drag and increased delta V imparted. Bank angle modulation is commanded throughout the atmospheric pass from point 3 to point 5. By point 5, where the influence of aerodynamic forces is no longer significant, the energy depleted from the initial hyperbolic trajectory is that required to capture the vehicle into the desired orbit. At apoapsis, point 7, a small delta V burn is performed to raise the periapsis.

AEROCAPTURE CORRIDOR

The aerocapture theoretical corridor is bounded by the full lift up trajectory and the full lift down trajectory for a nominal atmosphere and vehicle aerodynamics. (The theoretical corridor width is defined by the difference between the entry flight path angle corresponding to a full lift down trajectory and the entry flight path angle corresponding to a full lift up trajectory for a nominal atmosphere and aerodynamics.) If the vehicle enters the atmosphere at a flight path angle steeper than defined by the full lift up trajectory the vehicle lands. If the vehicle enters the atmosphere at a flight path angle shallower than that defined by the full down trajectory the vehicle is not captured. Figure 2 illustrates the theoretical corridor and the effects of navigation, atmosphere and aerodynamic uncertainties on the theoretical corridor. The plot shows entry flight path angle for a full lift up and full lift down trajectory as a function of the atmosphere variable Fminmax, where Fminmax=0 is the nominal atmosphere, Fminmax=-1 is the lowest density atmosphere, Fminmax =+1 is the maximum density atmosphere. The plot is shown for a given vehicle, entry velocity, and target orbit. To first order, the aerocapture corridor required to accommodate atmospheric dispersions, navigation errors (delivery flight path angle), and aerodynamic uncertainties can be root sum squared to determine the total corridor required. If the theoretical corridor is significantly greater than

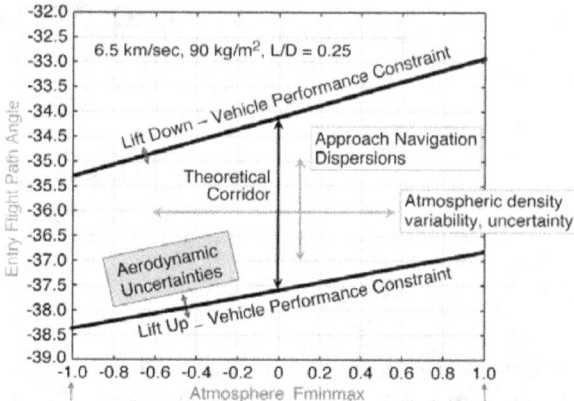

Figure 2. Aerocapture flight corridor, effect of uncertainties and dispersions.

the corridor width required, then the vehicle control authority is adequate and the aerocapture is robust. The approach is only an estimate. For example factors such as high frequency atmospheric density perturbations are not included in this approach and can affect the results. Monte Carlo simulation analyses must be completed to assess feasibility and robustness.

Concerns expressed regarding the risk of aerocapture have largely been in three areas. 1) Atmosphere variability and uncertainty; 2) Approach navigation delivery errors; 3) Aerodynamics utilized to control capture. To address these concerns, the following approach can be taken in the design of the aerocapture system.

- Based on available atmospheric measurements, quantify and model the physical range of atmosphere variability and uncertainty.
- Quantify the approach navigation delivery errors, and incorporate navigation systems into the vehicle design to reduce errors.
- Quantify the aerodynamic uncertainties including margin.
- Select a vehicle L/D to provide adequate control authority.
- Provide adequate vehicle control responsiveness to accommodate perturbations, including atmospheric perturbations.
- Develop a robust guidance.
- Evaluate aerocapture robustness through Monte Carlo simulation incorporating all variability, uncertainty, errors and dispersions.
- Design the aerocapture system to provide margin *above* 3-σ success, in particular for first time flights and for high value payloads.

PREVIOUS AEROCAPTURE ASSUMPTIONS

Previous aerocapture mission analysis work has demonstrated significant benefit of aerocapture for outer planet missions. However, due to the broad nature of the work, the analysis was low fidelity. Many assumptions were made, and it was noted that higher fidelity analyses would be needed to address the assumptions.

Assumptions made in previous aerocapture mission analyses were that the available aerocapture corridor width at a given destination was adequate to accommodate all of the dispersions and uncertainties, such as atmosphere uncertainties, navigation dispersions, and aerodynamic uncertainties. And that these uncertainties and dispersions could be quantified. It was assumed that guidance algorithms could be designed to successfully capture the vehicle over the range of uncertainties/dispersions. For Titan, the aerocapture subsystem mass, including structure and TPS, was estimated to be 27% of entry mass. It was assumed that aeroheating environments for aerocapture could be quantified and accommodated by TPS systems within the mass allocation. With entry velocities at Titan of 6-10 km/sec and the Titan atmosphere being predominantly Nitrogen, aeroheating rates were expected to be modest and convective heating dominated. As a result, TPS materials such as SLA 561 were expected to be applicable. All of the earlier studies assumed that the spacecraft could be packaged in the aeroshell; that volume is available, subsystems can be located to meet center of gravity (cg) restrictions, thermal and communication issues can be solved. It was assumed that the selected science orbits are feasible and that technology, including analysis tools would be ready in time to meet a project schedule. It was also estimated that the benefit of aerocapture at Titan compared to an all-propulsive capture provides a mass savings of 66%.

All of the assumptions made were to be addressed in the systems analysis study described here.

OBJECTIVES

The objectives for the Titan aerocapture systems analysis were therefore to provide higher fidelity analyses for validation and update to aerocapture assumptions made in mission studies, including performance, environments, mass properties, etc. The results of the analysis were to be provided to scientists, mission planners, technology planners, technologists and future mission managers. The feasibility, benefit and risk of aeroshell aerocapture system and technologies for Titan destination were to be defined. Technology gaps were to be identified and performance goals of key technologies defined.

APPROACH

A multi-center aerocapture systems analysis team was formed, including NASA engineers and scientists from Ames Research Center, the Jet Propulsion Laboratory, Johnson Space Center, Langley Research Center, and Marshall Space Flight Center, led by Langley Research Center. The team kicked off the study in late February and completed the work in September 2002.

The analysis included top level sensitivity studies to identify a reference concept for higher fidelity analysis and to provide sensitivities through a broader range of possible aerocapture mission scenarios. The reference concept was to provide a higher fidelity reference to address the previous assumptions noted above, to provide a reference for higher fidelity component level trades, and to provide a benchmark to the top-level sensitivities.

The mission objectives and initial spacecraft design for the reference concept was based on JPL's TeamX study[1] of the Titan Explorer mission.

TITAN AEROCAPTURE REFERENCE CONCEPT

The level one objectives for the Reference Concept were defined based on a modified set of those used in the TeamX study. The Titan Explorer consisted of an Orbiter and a Lander, each delivered to Titan. The Orbiter delivers the Lander to its Titan entry trajectory. The Lander performs a direct entry. The Orbiter aerocaptures into a near polar orbit about Titan.

One of the goals is to minimize trip time to Titan. The science mission is to be three years, with three years of Orbiter operations at Titan. In addition, the Orbiter serves as a relay for the Orbiter for the first year. The Orbiter science instruments include a Microwave spectrometer, SAR, Multispectral imager, USO as described in more detail in Ref 4. The launch date selected is 2010. This requires technologies to be at a technology readiness level (TRL) of six by 2006. Other launch dates are also considered. A SEP propulsion module is selected for the Reference Concept with comparisons made to the chemical propulsion module. Given the relatively near-term launch date, as much heritage hardware as possible is utilized. The mission is defined as a Class A mission with a fully redundant design. The team was funded to study the aerocapture Orbiter only. The Lander had to be treated as a "black box" such that most of the

TeamX analysis for the Lander was taken as is. The Lander mass allocation is 400 kg.

CONCEPT SELECTION

Selection of a Reference Concept is a balance between providing reduced trip time to the destination, adequate delivered mass, adequate aerocapture flight corridor width to provide a robust system, all while meeting the science objectives.

Mission analyses were conducted for the Titan aerocapture mission with various launch vehicles, gravity assist options, chemical propulsion vs. solar electric propulsion (SEP), and various launch dates as described in Reference 3. Figure 3 shows the delivered mass vs. flight time for various launch dates for a Venus Gravity Assist (VGA) and a SEP propulsion module. Note that for flight times greater than six years, the maximum mass that can be delivered remains nearly constant. Delivered mass decreases significantly as trip time is reduced from six to five years or less. Figure 4 shows the inertial entry velocity for the same missions as Figure 2. Six to eight year trip times result in entry velocities of less than ~7 km/sec for most launch opportunities. Also, entry velocity increases rapidly for trip times less than five to six years.

Since convective heating increases approximately with the cube of entry velocity as shown in Figure 4, trip times greater than six years were expected to reduce the mass of the required thermal protection system allowing more of the delivered mass to be allocated to meet science requirements.

Selection of a 6.5 km/sec entry velocity with an approximately 6 year trip time, dependent on opportunity, resulted in a balance between trip time and mass. From Figure 5, the stagnation point convective aeroheating on a one meter nose radius at 6.5 km/sec is approximately 40 W/cm^2. SLA 561 was expected to be adequate for the aeroheating environment.

Note that due to the Titan orbit about Saturn, entry velocity can be increased for given flight time without loss in delivery mass capability. The velocities in Figure 4 are each minimum inertial entry velocity.

To determine the theoretical corridor width leading to determination of the vehicle lift to drag ratio required, full lift up and full lift down aerocapture trajectories were developed over the range of entry velocities of interest, and for a range of lift to drag ratio vehicles. Initial analysis had shown that ballistic

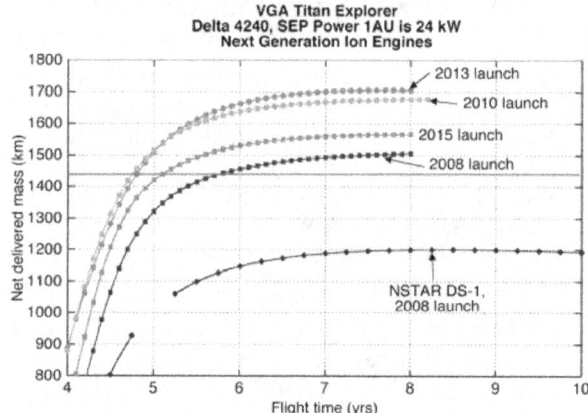

Figure 3. Delivered mass vs. mission flight time. Ref 3.

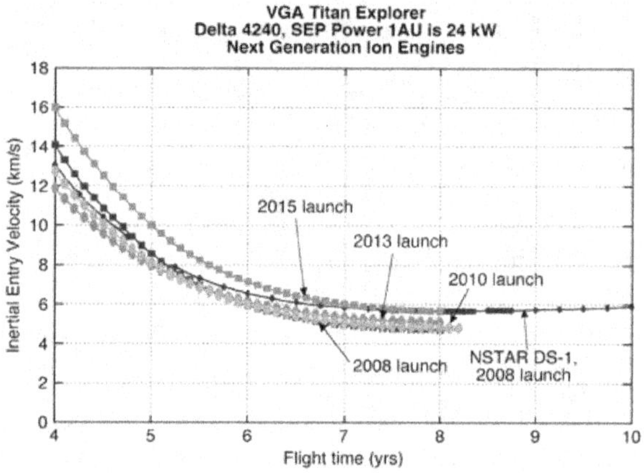

Figure 4. Inertial entry velocity vs. mission flight time. Ref 3.

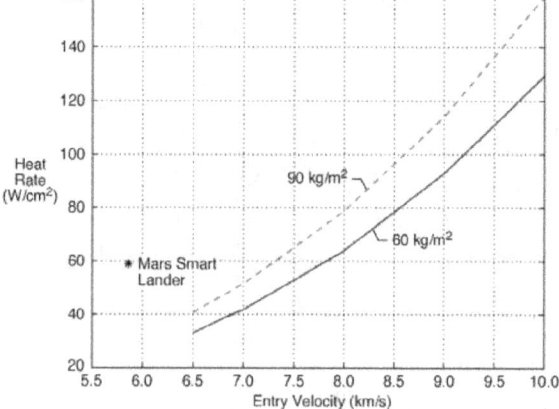

Figure 5. Convective stagnation point heat rate vs. entry velocity. Ref 7.

coefficient and target apoapsis were secondary variables in the determination of corridor width. From Figure 6, for a vehicle with L/D = 0.25, a 3.5 deg theoretical corridor width with 6.5 km/sec entry velocity is achieved. (With a 10 km/sec entry and the same L/D, 4.7 deg theoretical corridor width is achieved.) The 3.5 deg corridor width was expected to be more than adequate to accommodate 3-σ navigation delivery errors, atmosphere dispersions and aerodynamic uncertainties with 99.7% or greater success. This allowed a high heritage low L/D sphere cone configuration to be selected. If increased corridor were required, mid L/D configurations are viable alternatives.

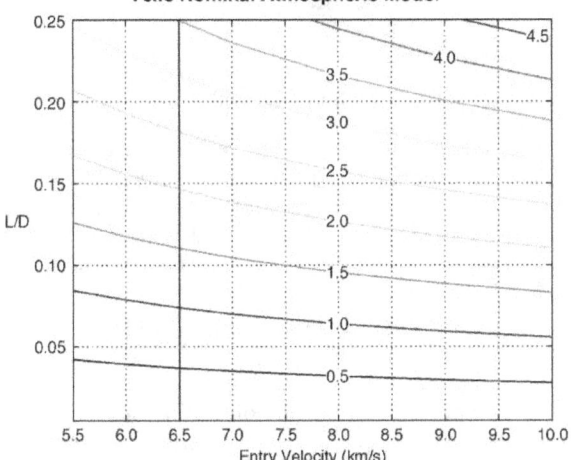

Figure 6. Aerocapture corridor width as a function of vehicle lift to drag ratio. Ref 7.

NAVIGATION

Several navigation approaches were analyzed for Titan as described in Reference 2. The approach selected for the Reference Concept is based on the Cassini optical navigation camera capability, ΔDOR (utilized on Odyssey), and Dopler and Range combined. The 3-σ flight path angle dispersions, based on a Beginning of Cassini (BOC) Titan ephemeris is ±1.42°. Significant improvements in knowledge of the location of Saturn, Titan will result from Cassini mission. 3-σ sets of entry states, BOC and EOC, were utilized in the Monte Carlo analyses described below. This detailed navigation analysis was required to assess feasibility and robustness of Titan aerocapture. Further reduction in the flight path angle dispersions can be achieved with the resolution of the Mars Reconnaissance Orbiter camera, as noted in Reference 2.

ATMOSPHERE MODELING

TitanGRAM, discussed in detail in Reference 5, includes a model of measurement uncertainties, residual uncertainties (turbulence, waves, etc); variation with latitude, altitude, time of day, season. This model fidelity is required to assess mission feasibility and robustness. Figure 7 shows the Titan atmospheric density with altitude.

The arrival date of the current study results in the maximum variation of density with latitude. Since the science orbit is near polar, aerocapture occurs over a wide range of latitudes. Figure 8 represents a simulation of the expected variability of Fminmax with latitude.

Figures 7 and 8 show the mean variation of the Titan atmosphere. Also included in TitanGRAM are the atmosphere perturbations. Both the mean variations and perturbations are utilized in the Monte Carlo analysis described below.

Cassini-Huygens data will reduce measurement uncertainty as discussed further below.

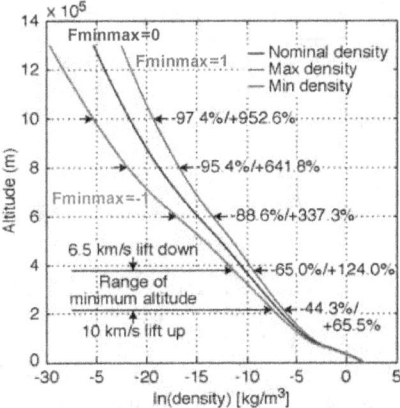

Figure 7. TitanGRAM atmospheric density as a function of altitude.

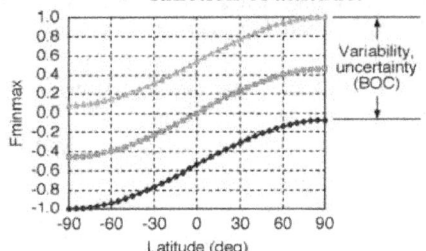

Figure 8. Simulated variation of Fminmax TitanGRAM parameter with Titan latitude.

AERODYNAMICS

Aerodynamic uncertainties included in the Monte Carlo are based on LAURA CFD for blunt bodies at hypersonic velocities.[7] The uncertainties include ±3% uncertainty in CA; ±5% uncertainty in CN; and a trim angle of attack (α_{trim}) uncertainty of ±2 deg used to represent uncertainty in Cm. The CG uncertainty used is ±0.0318 m in axial, Zcg and ±0.0069 m in radial Xcg. Figure 9 (Ref 7) shows the effect of these uncertainties on vehicle lift to drag ratio.

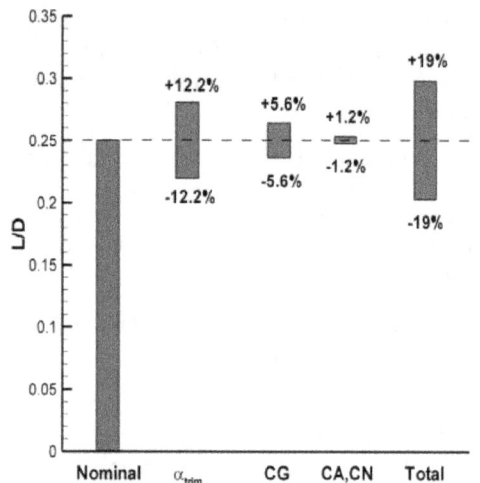

Figure 9. Effect of aerodynamic uncertainties on L/D. Ref 7.

GUIDANCE

HYPAS guidance (ref 6) was chosen for the Titan aerocapture systems analysis. (Other guidance algorithms are planned for future consideration.) HYPAS was originally developed for the Aeroassist Flight Experiment (AFE). However the current version includes several improvements since AFE.

HYPAS utilizes vehicle lift and bank angle control through the atmosphere to target the desired exit orbit apoapsis and inclination (or plane). It is an analytically derived algorithm based on deceleration due to drag and altitude rate error feedback. This analytic, non-iterative, on-the-fly approach leads to efficient code (~320 source lines in Fortran), minimal storage requirements, and fast and consistent execution times.

HYPAS consists of two phases as illustrated in Figure 10: 1) Capture Phase: Establishes pseudo-equilibrium glide conditions; 2) Exit Phase: Exit conditions are predicted analytically assuming a

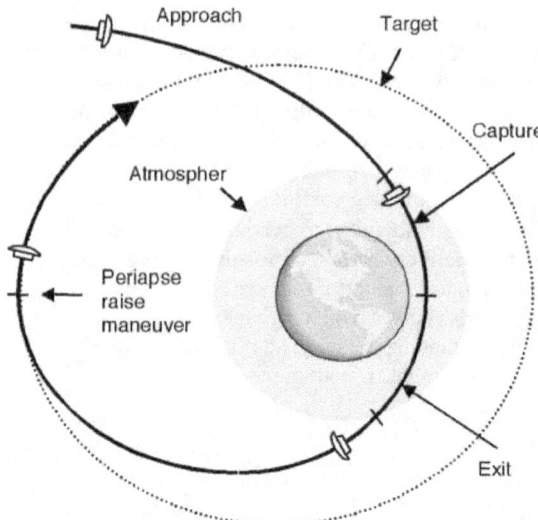

Figure 10. Aerocapture guidance phases. Ref 6.

constant altitude rate. The lift vector is adjusted to null the error between predicted and target apoapsis, and bank reversals are used to keep inclination errors within the desired limits. Results show excellent performance and an ability to capture ~98% of the theoretical corridor.

PERFORMANCE/SIMULATION

A Titan aerocapture simulation and Monte Carlo have been developed. All uncertainties critical to aerocapture robustness are included in the parameters varied in the Monte Carlo simulations. These parameters are the initial states with corresponding delivery errors at atmospheric interface; the atmosphere variability and uncertainty; including mean range and perturbations; aerodynamic uncertainties in CA, CN, α_{trim}, and cg uncertainties.

During a Monte Carlo analysis, Monte Carlo parameters are each randomly varied over a specified range and distribution defined by the subsystem/model engineer. Two thousand trajectory simulations are completed in one Monte Carlo analysis. Results from these simulations provide aerocapture performance statistics to determine robustness, margin, risk; guidance development, stress case identification (control algorithm development – future); statistical distributions of critical parameters, design trajectories, for subsystem design.

Monte Carlo analyses are completed for six Titan aerocapture mission scenarios. The first three are completed for the inertial arrival velocity of 6.5 km/sec with no updates assumed from

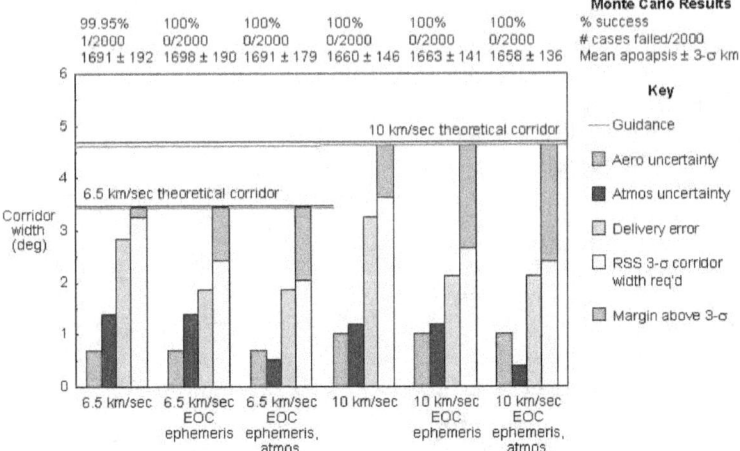

Figure 11. Titan aerocapture robustness.

Cassini/Huygens. This analysis is labeled "6.5 km/sec" on the chart in Figure 11. As shown at the top of the figure, for this Monte Carlo 99.95% of the cases successfully aerocaptured at Titan; one of 2000 cases failed. The first set of bars in Figure 11 show the corridor width required to accommodate aero uncertainty, atmosphere uncertainty, and delivery error, as well as the root sum square used to estimate total corridor width required for the 6.5 km/sec case. A comparison of the corridor width required and the theoretical corridor minus the loss in corridor due to guidance (only 2% loss), indicates that the 6.5 km/sec case with no update from Cassini/Huygens will be successful 3-sigma, but that there is not significant margin over and above a 3-sigma success.

Adding the expected improvement in ephemeris from Cassini/Huygens in the next Monte Carlo analysis, labeled "6.5km/sec EOC ephemeris" shows the reduction in corridor required to accommodate delivery errors, and a corresponding reduction in the RSS 3-sigma corridor width required. The result is a significant increase in margin over and above a 3-sigma success. These results are confirmed by the Monte Carlo analyses, where 100%, or 2000 of 2000, cases successfully aerocaptured. Further improvement results from the expected reduction in atmospheric uncertainty resulting from the Cassini/Huygens mission, labeled "6.5km/sec EOC ephemeris, atmos."

The next set of Monte Carlo's completed is for 10km/sec inertial entry velocity. The same three scenarios are completed for the 10km/sec entries as for the 6.5 km/sec entries. For each scenario, 100% of the Monte Carlo cases are successfully captured. Note that with increased velocity, the theoretical corridor increases faster than the net change in corridor width required due to aero, delivery and atmosphere errors and dispersions. Thus aerocapture performance robustness increases with velocity. The simulation, Monte Carlo analysis and results are discussed further in reference 7.

AEROHEATING ENVIRONMENTS

Conservative aeroheating design trajectories were selected for the reference concept. Design trajectories were based on the maximum atmospheric concentration of CH_4, 5% by volume. The atmospheric CH_4 concentration estimate is expected to either remain or be decreased with Cassini/Huygens data.

A conservative ballistic coefficient, 90 kg/m^2, was assumed for the design trajectories. The resulting reference concept ballistic coefficient is 69kg/m^2, with a possible range of 56 – 84 kg/m^2 for variation in vehicle diameter, aeroheating rates/loads, TPS selection and sizing (shown later in the paper). In addition, initial analyses show a decrease in aeroheating with decreasing ballistic coefficient, making 90 kg/m^2 conservative for the design trajectories.

Lift-up trajectories were utilized to define the maximum heat rates; lift-down trajectories were used to define the maximum heat loads. These selections result in the most conservative trajectories for aeroheating environment definition.

Based on the design trajectories and vehicle configuration, the peak laminar convective aeroheating rates are ~46 W/cm^2 for 6.5 km/sec lift up, minimum atmosphere, alpha = 16 deg, stagnation point. Transition to turbulence is likely prior to peak convective heating on heatshield lee side based on an Retheta = 200 transition criteria and CFD results for the 6.5 km/sec lift up, minimum atmosphere, 16 deg angle-of-attack trajectory. This will likely increase the maximum convective heating on the heatshield beyond the above 46 W/cm^2.[9]

For the maximum Titan CH4 concentrations, 5% by volume, all aeroheating predictions from the study have shown that the radiative aeroheating from CN is greater than the convective aeroheating. For the windside of the heatshield at alpha = 16 deg, the maximum radiative aeroheating is predicted to occur at the stagnation point. Radiative heating rates range from the "low end" ~93 W/cm^2 to the "conservative" ~280 W/cm^2 for 6.5 km/sec lift up, minimum atmosphere, alpha = 16 deg, stagnation point. These results, completed during the study, were known to be preliminary. One of the recommendations from the study was to develop improvements in the radiative aeroheating environments analysis methods. This work is currently on going and discussed further in References 8 and 9.

Based on the preliminary estimates, a range of radiative and convective heat rates and loads are estimated to provide a sensitivity of TPS selection and sizing to potential aeroheating environments for Titan aerocapture as shown in Figure 12.

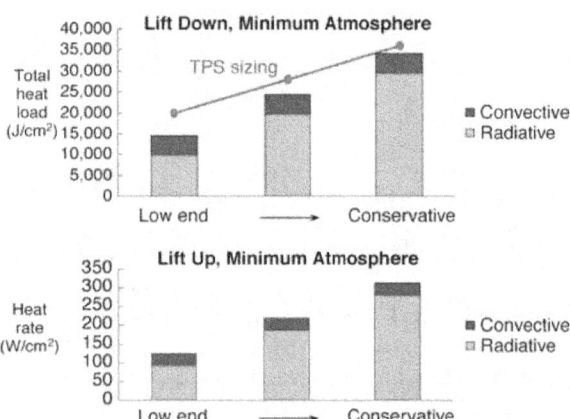

Figure 12. Aeroheating environments.

TPS

Candidate TPS materials are identified, and several sized,[10,11,12,13] for the range of Titan aeroheating environments expected. The ability of low density TPS materials (ex. SRAMs) to absorb radiation is currently unknown and requires testing. TUFROC, PhenCarb20, C-C Genesis-type are expected to absorb radiation. However these materials must also

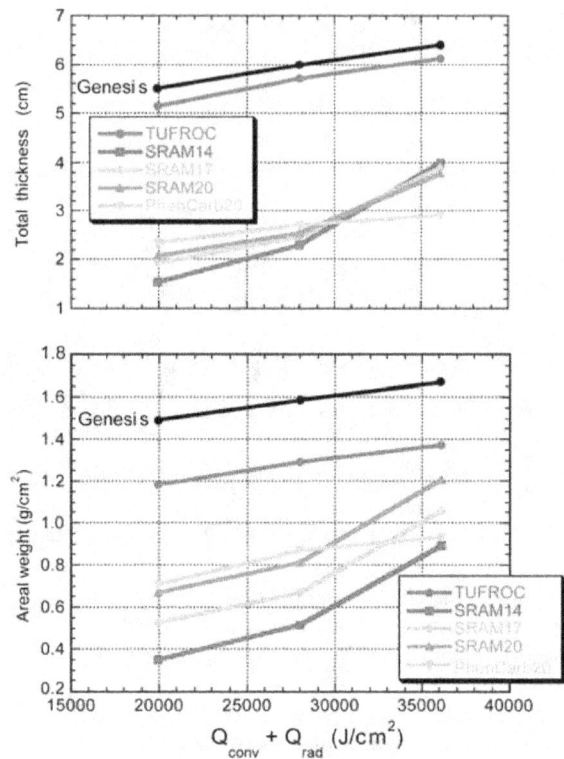

Figure 13. TPS thickness and mass vs. TPS type, aeroheating load. Ref 10.

be tested in the radiative environments. In the TPS sizing analyses shown in Figure 13,[10] it was assumed that the materials could absorb the radiation.

Low-density silicone-based ablators are predicted to experience significant recession at the highest heating conditions. However they provide the lowest mass at lowest heating conditions. TUFROC and PhenCarb20 exhibit less sensitivity to heating variations than low density silicone-based ablators, but with increased mass. PhenCarb 20 exhibits small surface recession in the inert Titan atmosphere. TUFROC is non-ablating. C-C foam Genesis-type concept is also non-ablating.

For the Reference Concept, TUFROC (with the low end heat load sizing + 30%) was utilized in the MEL for two reasons. The TUFROC is expected to absorb radiation, and TUFROC TPS results in a conservative mass estimate compared to other candidate materials.

TPS testing is needed to determine the ability of TPS materials to absorb CN radiation and to provide a family of TPS materials to accommodate a range of potential aeroheating environments at reduced mass.

AEROSHELL STRUCTURE

The aeroshell structure was designed to be a current technology concept.[14] Figure 14[14] illustrates the structural components of the aeroshell. The loads are launch dominant with launch loads of 7 G axial, 3 G lat; and entry loads of 4 G axial with 3146 Pa on the heatshield for the 6.5 km/s entry. TPS masses used in the structure design are 1.181 g/cm^2 for the heatshield TUFROC TPS, and .187 g/cm^2 for the backshell SLA TPS.

The aeroshell was sized using HyperSizer/NASTRAN with the following results. The heatshield is 25.4 mm thick Hexcell 5052 Alloy Hexagonal Al Honeycomb core, with Graphite Polyimide facesheets. The backshell is 12.7 mm thick Hexcell 5052 Alloy Hexagonal Al Honeycomb core, with Graphite Polyimide facesheets.

The aeroshell mass is summarized in Figure 15.[14] The total aeroshell current best estimate mass (structure + TPS + non structural) is 327.80 kg.

MASS PROPERTIES, PACKAGING

Figure 16 shows the stack and orbiter packaging design.[4] Note that the aeroshell size and packaging efficiency are governed by the 2.4 m diameter high gain antennae (HGA) packaging.

Figure 17 shows a system level mass summary for the Delta 4450, SEP, Earth Gravity Assist (EGA), aerocapture concept. The concept has 30% system level margin, and greater than 10% system reserve. A VGA option with a Delta 4450, SEP, VGA, aerocapture has 6% system reserve. The aerocapture mass fraction is 41.5% of orbiter entry mass. These results are not possible without this level of detail in packaging, s/c design, structure, and TPS.

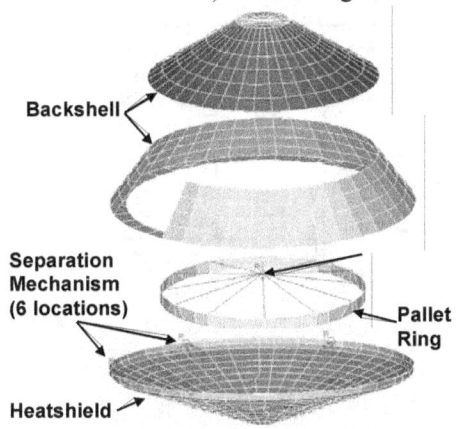

Figure 14. Aeroshell structural components. Ref 14.

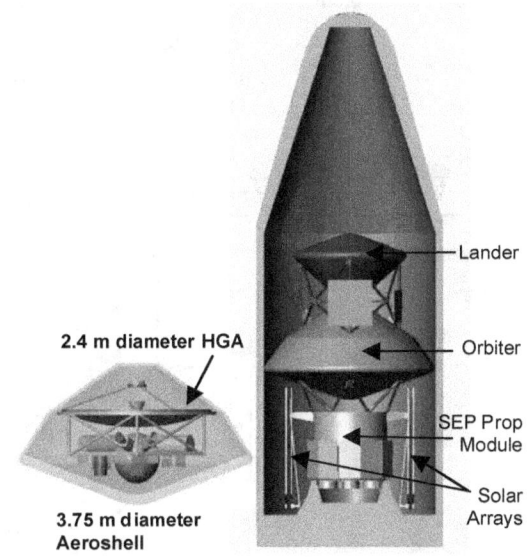

Figure 16. Refererence mission packaging. Ref 4.

Component	Area (m^2)	Structure Mass (kg)	TPS Mass (kg)	Non Structural Mass (kg)
Heatshield	12.58	41.58	148.62*	0
Backshell	15.01	43.27	28.69**	2.38
Pallet Ring	1.20	42.47	0	1.20
Sep Ring	1.79	11.35	0	.89
Sep Ring Attachments	0.45	2.85	0	4.50
Totals		141.52	177.31	8.97

Figure 15. Aeroshell structure mass. Ref 14.

	Mass (kg) Subsystem Rack-up			
Component	Current Best Estimate	% Contingency	Growth	System Allocation
Lander	280.2	29.8%	363.8	400.0
Orbiter/Lander Interface	47.5	30.0%	61.8	61.8
Orbiter	883.6	24.2%	1097.7	1200.0
Prop Mod/Orbiter Interface	47.3	30.0%	61.4	61.4
SEP Prop Module	1084.0	21.4%	1316.5	1450.0
Launch/Prop Mod Interface	60.0	30.0%	78.0	78.0
Stack Total	2402.6	24.0%	2979.2	3251.2
Launch Vehicle Capability	3423			
System Level Mass Margin	29.8%	(LV Cap - CBE) / LV Cap		
System Reserve	13.0%	(LV Cap - Growth) / LV Cap		

Figure 17. System level mass properties for Reference Titan Explorer aerocapture mission. Ref 4.

MASS PROPERTY SENSITIVITIES

Several mass property sensitivity analyses were completed to determine aerocapture mass fraction, aeroshell mass, system mass sensitivity to aeroheating environment assumptions, TPS candidates and aeroshell size.[15] An assessment of aerocapture system mass to an all propulsive mass system was also completed to determine the mass savings for an aerocapture system.[15, 3]

The aerocapture system mass fraction as a function of aeroheating environment and TPS concept is shown in Figure 18.[15] For example, SRAM-20 is sized for three levels of aeroheating environments. Results are labeled SRAM-20, SRAM-20 +1, SRAM-20 +2, for the low, medium and conservative aeroheating levels respectively. The potential savings in mass for alternate TPS concepts, compared to the baseline, is evident. In addition, the effect of aeroheating environment on TPS mass is also shown. The ballistic coefficient range is 56 – 72 kg/m^2 over this range of aeroheating environments and TPS materials for the 3.75 m diameter aeroshell.

Based on the orbiter packaging, a reduction in the aeroshell diameter may be possible. The minimum diameter would be approximately 3 m. Further reduction would preclude packaging of the spacecraft with the 2.4 m HGA. A maximum of 20% reduction in aerocapture system mass fraction results for a minimum 3 m diameter aeroshell. The ballistic coefficient range for aeroshell size range of 3 – 3.75 m diameter and range of TPS is 56 – 84 kg/m^2, all lower than the design trajectory ballistic coefficient of 90 kg/m^2.

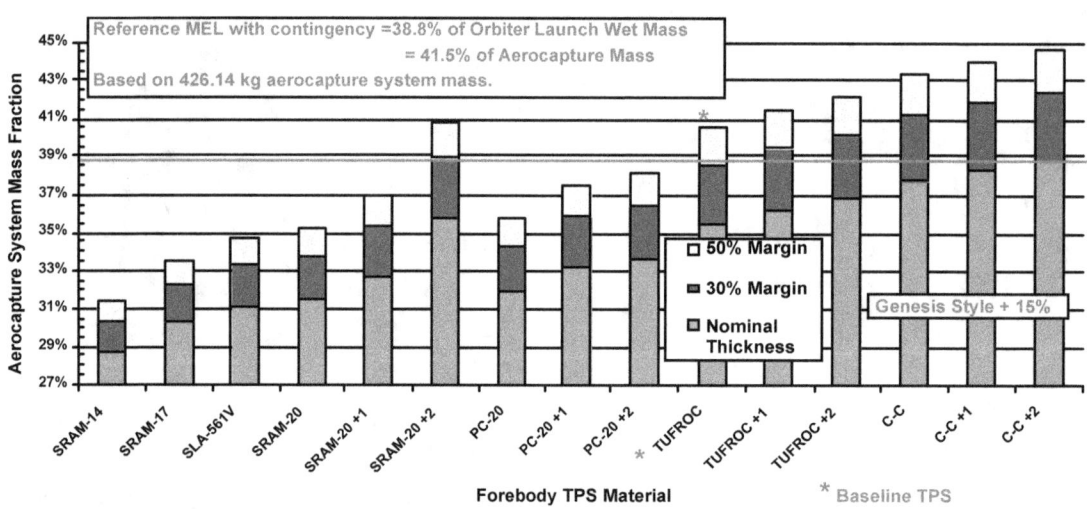

Figure 18 Aerocapture system mass fractions vs. forebody TPS material and aeroheating environments.

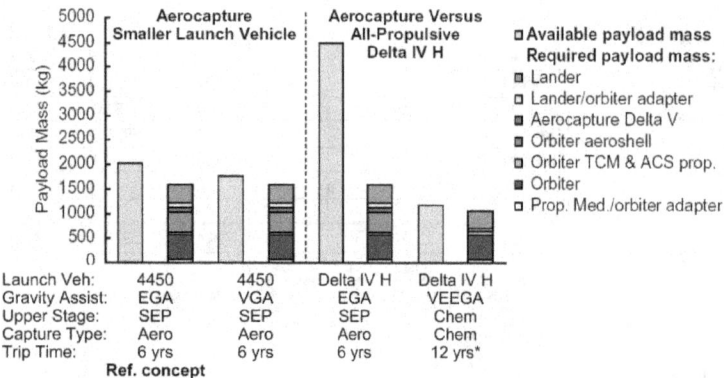

Figure 19 Aerocapture vs. All-Propulsive. *Includes 2 year moon tour used to reduce the propellant requirements for all propulsive capture.

Figure 19 shows a comparison of the Titan Explorer mission with aerocapture compared to an all-propulsive option. For the Titan Explorer payload mass, the all-propulsive option was not feasible with a Delta 4450, while aerocapture enables a feasible mission on the 4450. On a Delta IV Heavy an all-propulsive option is marginal, but potentially feasible. However trip times are 12 years for an all propulsive option compared to 6 years for an aerocapture mission. In addition the aerocapture mission is capable of delivering 2.4 times more payload to Titan compared to the all-propulsive mission for the same launch vehicle. Dependent on mission requirements including required payload mass and trip time, for example, aerocapture is enabling to strongly enhancing compared to an all-propulsive mission.

SUMMARY

Aerocapture is feasible, and the performance is adequate, for the Titan mission with the high heritage low L/D configuration and no improvements from the Cassini/Huygens mission. 99.95% success results for a 6.5 km/sec entry with the low L/D configuration, but with low margin above 3-σ. 100% success results for a 10 km/sec entry with the low L/D configuration, with margin above 3-σ success. Cassini/Huygens results in an increase in margin and robustness significantly above 3-σ, for 6.5 km/sec and 10 km/sec entry aerocapture at Titan with the high heritage low L/D configuration.

Aerocapture can deliver 2.4 times more mass to Titan than an all-propulsive system for the same launch vehicle.

TECHNOLOGIES

Technologies identified in the study as needing development were grouped into three categories; enabling technologies, strongly enhancing technologies and enhancing technologies. As noted earlier, aerocapture is enabling to strongly enhancing for the Titan missions, dependent on the mission requirements. However aerocapture is ready for an Earth or Mars flight experiment without additional technology development. An aerocapture flight experiment may be desirable to demonstrate aerocapture technology to reduce risk/cost for science mission acceptance.

The strongly enhancing technologies identified include:

- Aeroheating methods development and validation. Large uncertainties currently exist, improved prediction capability could result in reduced TPS mass, as well as support configuration trades and further mission design trades.

- TPS material testing. TPS materials proposed and other TPS options exist today, but have not been tested against the expected radiative heating at Titan.

- Atmosphere Modeling. Titan General Circulation Model output is needed to represent the "true" natural variability of the atmosphere.

- The enhancing technologies identified include

- Aeroshell lightweight structures for reduced aerocapture mass.

- Guidance – Existing guidance algorithms have been demonstrated to provide acceptable performance, improvements could provide increased robustness.

- Simulation – Huygens trajectory reconstruction, statistics and modeling upgrades.

- Mass properties/structures tool for systems analysis capability improvement, concept trades.

- Deployable high gain antennae for increased data return.

- The following technologies provide significant benefit to the mission but are already in a funded development cycle for TRL 6 by 2006.

- MMRTG (JPL sponsored AO in proposal phase, First flight Mars '09)

- SEP engine (Glenn Research Center engine development complete in '0#)

- Second Generation AEC-Able UltraFlex Solar Arrays (175W/kg)

- MRO optical navigation camera to be demonstrated in MRO.

ADDENDUM

Aerocapture Characteristics: Titan vs. Mars

Since much of the higher fidelity aerocapture work has been done for Mars missions, Figure 20 provides a comparison of a representative Mars aerocapture mission to the Titan aerocapture reference mission described in this paper for reference.

Destination/ Reference Parameter	Titan	Mars
Entry Velocity (km/sec)	6.5*	5.7
Nom. Entry Flight Path Angle (deg)	-36	-14.2
Apoapsis/Science Orbit (km)	1700	1400
Atmosphere Composition	95%N2, 5%CH4**	95.3% CO2, 2.7%N2
Atmos Scale Height at Aerocapture Altitude	~40	10.5
Atmospheric Interface Altitude (km)	1000	250
Aerocapture Altitude (km)	200-400	40
L/D	.25	.25
M/CDA (kg/m^2)	90***	148
Theoretical Corridor (deg)	3.5	~1.4
Time from Atmos Entry to Atmos Exit (min)	42	10
Convective Stagnation Point Heat Rate (W/cm^2)	46 (.91m nose rad)	30 (1.9m nose radius)
Radiative Aeroheating Rate (W/cm^2)	93-280	negligible
Max g's During Aerocapture (Earth g's)	3.5	2.5 - 3

Figure 20. Representative aerocaptures at Titan compared to Mars. *Titan aerocapture entry velocity 6.5 – 10km/sec, comparison given for 6.5km/sec. **Max CH4 atmosphere. ***For design trajectory & comparison; range: 56–84 kg/m^2 dependent on aeroheating, TPS, vehicle diameter

REFERENCES

1. "Titan Lander Conservative Science 01-06" results dated 4,5,8,15 June, 2001.

2. Haw, R. "Approach Navigation for a Titan Aerocapture Orbiter," AIAA-2003-4802.

3. Noca, M., Bailey, R., "Titan Explorer Mission Trades from the Perspective of Aerocapture,"AIAA-2003-4801.

4. Bailey, R., Hall, J., Spilker, T., "Titan Aerocapture Mission and Spacecraft Design Overview," AIAA-2003-4800.

5. Justus, J., Duvall, A., "Engineering-Level Model Atmospheres for Titan and Neptune", AIAA-2003-4803.

6. Masciarelli, J., "Guidance Algorithms for Aerocapture at Titan," AIAA-2003-4804.

7. Way, D., Powell, R., Edquist, K., "Aerocapture Simulation and Performance for the Titan Explorer Mission," AIAA-2003-4951.

8. Olejniczak, J., Prabhu, D., Wright, M., Takashima, N., Sutton, K., "An Analysis of the Radiative Heating Environment for Aerocapture at Titan," AIAA-2003-4953.

9. Takashima, N., Hollis, B., Olejniczak, J., Wright, M., Sutton, K., "Preliminary Aerothermodynamics of Titan Aerocapture Aeroshell," AIAA-2003-4952.

10. Laub, B., "Thermal Protection Concepts and Issues for Aerocapture at Titan," AIAA-2003-4954.

11. Curry, D.M., Congdon, W.M., "2007 Titan Mission Analysis of ARA Ablators to Define Required Thickness and Weights," Report presented to Titan Aerocapture Systems Analysis Team, Report Date 11July2002.

12. Congdon, W.M., "TPS-Table-Titan," e-mail to M.K.Lockwood, 23Aug2002.

13. Dec, J.A., "Titan Orbiter Heatshield Sizing Study," e-mail to M.K.Lockwood, 27Sep2002.

14. Hrinda, G., "Structural Design of the Titan Aerocapture Mission," AIAA-2003-4955.

15. Dyke, R. E., "Planetary Probe Mass Estimation Tool Development and Its Application to Titan Aerocapture," AIAA-2003-4956.

Acknowledgements: The author would like to acknowledge and thank the team members of the NASA Aerocapture Systems Analysis Team for their work and contributions to the Titan Aerocapture Systems Analysis Study and to this paper. Thank you to Carol Carroll, Aerocapture Systems Analysis Study Review Panel Chair, and the review panel members for review of this work and helpful comments and recommendations. Thank you to Anne Costa for preparing this paper for publication.

TITAN AEROCAPTURE MISSION AND SPACECRAFT DESIGN OVERVIEW

R. W. Bailey, J. L. Hall, T. R. Spilker
Jet Propulsion Laboratory
California Institute of Technology

ABSTRACT

A detailed Titan aerocapture systems analysis and spacecraft design study was performed as part of NASA's In-Space Propulsion Program. The primary objective was to engineer a point design based on blunt body aeroshell technology and quantitatively assess feasibility and performance. This paper provides an overview of the mission and spacecraft design resulting from that study and references other papers that provide further details on critical subsystems. It also reviews the science requirements underlying the selected mission concept of an aerocaptured orbiter and a separate entry vehicle that delivers an aerobot into the Titan atmosphere. Including aeroshells and 30% contingencies, the estimated mass of the orbiter is ~1100 kg and that of the entry vehicle ~360 kg. Solar electric propulsion (SEP) and an Earth gravity assist is used to get the tandem vehicle to Titan in 6.5 years, with orbiter – entry vehicle separation occurring one month prior to arrival. The SEP module, orbiter and entry vehicle are vertically stacked on a medium class launch vehicle and connected with a truss structure. Power profiles based on a strawman instrument suite and telecom strategy are accommodated with a pair of 120 W (electric) radioisotope thermoelectric generators. Details on the configuration layout, mass and power breakdowns, key design trades and outstanding design issues are also included.

TABLE OF CONTENTS

1. Introduction
2. Science
3. Mission Overview
4. Mission System Description
5. Atmospheric Probe Design
6. SEP Module Design
7. Orbiter Design
8. New Technology Development
9. Recommended Additional Analysis
10. Conclusions
11. Acknowledgements
12. Bibliography

INTRODUCTION

As part of NASA's In-Space Propulsion Program, aerocapture is being investigated as a means for interplanetary orbit insertion. A systems analysis and spacecraft point design study was performed in the Fiscal Year 2002 time frame based on a reference mission to Saturn's moon Titan. The purpose of this study was to quantify the feasibility and performance of an aerocapture system to insert a spacecraft into a science orbit about Titan. This paper provides an overview of the mission and spacecraft design resulting from that study and references other papers presented at this conference that provide further details on mission design, navigation, critical subsystems and the aerothermal environment for aerocapture at Titan.

The overall mission concept includes the delivery of a long duration atmospheric probe to Titan's atmosphere and the use of a Solar Electric Propulsion (SEP) stage for the Earth to Saturn transit. The mission concept is shown to be feasible at the level of detail applied for this study. Many different technical areas and trades consistent with continued Phase A/B efforts are defined at the end of this paper.

SCIENCE

Objectives & Measurements

For this study, the primary science objectives were taken from Chyba et al.[1] for a post-Cassini / Huygens Titan mission, listed in priority order:

1. Distribution and composition of organics
2. Organic chemical processes, their chemical context and energy sources
3. Prebiological or protobiological chemistry
4. Geological and geophysical processes and evolution
5. Atmospheric dynamics and meteorology
6. Seasonal variations and interactions of the atmosphere and surface (not addressed in a mission of short lifetime)

These objectives will likely be revisited when results are available from the Cassini / Huygens mission. Tamppari et al.[2] involved the Titan science community in a workshop that prioritized measurement objectives for such a mission. The highest priorities were determined to be:

1. Global surface morphology
2. Global gross surface composition and chemistry
3. Atmospheric composition and its spatial and temporal variability
4. Atmospheric structure and its spatial and temporal variability: vertical profiles of density, pressure, and temperature
5. Atmospheric dynamics (winds) and meteorology

An independent external review performed after the completion of this study judged these objectives to be appropriate.

Science Instruments

Table 1 presents the instrument suite selected and the flowdown from science and measurement objectives to the instruments. Although these instruments may be realistic, for the purpose of this study they serve as mass, power, and data volume placeholders for A Titan science payload.

Multi-Spectral Imager

The multi-spectral imager uses spectral coverage in several atmospheric opacity "windows" between 1 and 5 microns to determine surface and atmospheric morphology and chemistry as well as atmospheric dynamics and meteorology. This instrument will fill in any coverage gaps remaining after Cassini / Huygens.

Synthetic Aperture Radar (SAR)

The SAR uses the Orbiter X-Band telecom system with the HGA pointed off-nadir. It makes complementary measurements of surface morphology and meteorology, through clouds that would obscure the imaging instruments' view, and can detect the bottoms of shallow hydrocarbon lakes. Like the imager, it will fill in any coverage gaps remaining after Cassini / Huygens.

Microwave Spectrometer

The microwave spectrometer, capable of either nadir- or limb-pointed modes, makes global, low (spatial) resolution measurements of atmospheric structure, dynamics, and meteorology via detailed spectroscopy of emission lines from a few key chemical species. This also yields precise vertical abundance profiles of those species.

Ultrastable Oscillator (USO)

Adding a USO to the Orbiter X-Band telecom system enables atmospheric radio occultation science. Radiometrics obtained when the signal path to Earth passes through Titan's atmosphere allows accurate (1-2%), high-resolution vertical profiling of temperatures and densities at many sites, yielding atmospheric structure and dynamics as well as ionospheric structure

Atmospheric Probe Science

The Atmospheric Probe (AP) was allocated 5.3 Gbits of total data return; or the capability of the UHF relay link over a one year period. The AP to Orbiter link provides 64 kbps for 30 minutes every 8 days. Although 5.3 Gbits is adequate for general atmospheric and meteorological data (~14 Mbits/day), this volume is likely inadequate for any type of context imaging – this is generally an issue for the AP and not addressed in this study which focuses on the aerocapture technology aspects of the mission.

Table 1. Science Instruments

Instrument	CBE Mass (kg)	CBE Power (W)	Spatial Res (meters)	Point	FOV (deg)	Measure Objective	Science Objective	Total Data Return (Tbits)
Multi-spectral Imager	12	14	~30	Nadir	1.0	1,2,3,5	1,2,3,5	9
Synthetic Aperture Radar	10	30	~200	Off-Nadir	1.15	1,2	4	1.9
Microwave Spectrometer	10	50	N/A	Nadir & Limb	6.5	3,4,5	2,3,5	1.0E-4
Ultra Stable Oscillator	0.8	3	N/A	N/A	N/A	4,5	5	N/A

MISSION OVERVIEW

The study was based on a Titan Explorer concept with an Orbiter and an Atmospheric Probe (AP). Certain aspects of the mission were assumed as ground rules from previous studies performed internally at the Jet Propulsion Laboratory's (JPL) Team-X[12]. Other aspects of the mission were open to system trades and/or inherited from other outer planet mission studies performed internally at JPL.

Ground Rules

Several ground rules and assumptions were set to bound the study. These items were not subject to any system trades analysis.

- The mission shall deliver an AP into the Titan atmosphere, and a spacecraft into Titan orbit.
- The total mission lifetime shall be no longer than 10 years.
- The Technology Readiness Level 6 cutoff date shall be no later than Dec 2006.
- The AP will be a "black box" with a 400 kg launch mass allocation.
- The AP operational lifetime will be 1 year.
- The Orbiter shall perform an aerocapture for Titan orbit insertion.
- The Orbiter shall provide global coverage opportunity for all the science instruments.
- Science data return shall utilize no more than 8 hours per day of a 70m ground station.

Earth to Saturn Trajectory

The Earth to Saturn trajectory, shown in Figure 1, provides a good combination of transit time, Titan entry velocity, launch mass, and SEP propellant/power mass. Many different trajectories were considered which included different launch vehicles, launch dates, transit times, SEP power levels, number of SEP ion engines, and planetary gravity assists. These trajectory options and their associated trades are discussed in detail by Noca, et al.[3] The important aspects of the selected trajectory are as follows:

Launch Vehicle:	Delta IV M (4450-14)
Launch C3:	8.6 km^2/sec^2
Launch Mass:	3423 kg (10% reserve)
Launch Date:	Dec 24, 2010
Gravity Assist:	Earth
SEP Burn Time:	30 months, accumulated
SEP Power:	24 kW (End Of Life)
SEP Propellant:	460 kg (no contingency)
Transit Time:	5.9 years
Titan Entry Velocity:	6.5 km/sec

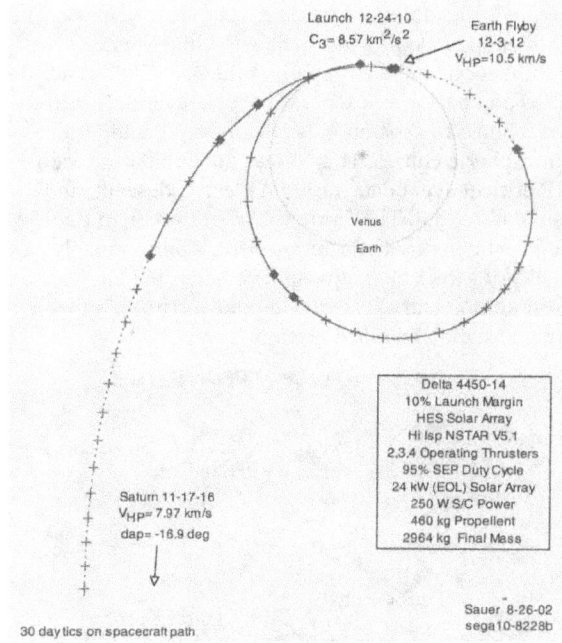

Figure 1. Earth to Saturn SEP trajectory

Mission Timeline

The mission timeline is listed below. For the "Time" column, 'L' = Launch, 'A' = Orbiter atmospheric interface, 'y' = years, 'd' = days, 'h' = hours, and 'm' = minutes.

Time	Event
L+0	Launch, SEP burn start
L+23m	Earth flyby
L+30m	SEP burn out and jettison at ~2.5 AU
L+5.7y	(A-60d) Traj Correction Maneuver (TCM) 1
A-31d	Probe Release TCM (2)
A-30d	Probe Release
A-29d	Post Release TCM (3)
A-7d	TCM 4
A-1d	TCM 5
A-6h	TCM 6 (if needed)
A-3h	Probe entry
A-1h	Jettison non-aero external components
A-30m	Align for aerocapture interface
A+20m	Jettison aeroshell
A+4h	Periapsis raise (circularization) burn
A+3y	End of mission

Once the Solar Electric Propulsion Module's (SEPM) job is done at around 2.5 AU, it is jettisoned to eliminate mass and solar array perturbations for later TCMs. The Orbiter uses a combination of Doppler ranging, ΔDOR, and optical navigation[4] to setup the AP entry trajectory delivery at entry minus 30 days.

Figure 2 illustrates the final aerocapture trajectory. The Orbiter spins up providing the AP with attitude stabilization, and then separates the AP. The Orbiter de-spins and performs a separation maneuver designed to put 3 hours of separation between the AP and Orbiter atmospheric entries. This allows the Orbiter to receive AP critical event data during AP entry, descent, and initial checkout. The Orbiter relays this data to Earth before it enters Titan's atmosphere. Approximately 1 hour prior to Orbiter atmospheric entry, the Orbiter will eject all non entry system components (truss, radiators, antennas, etc), and then orient for entry.

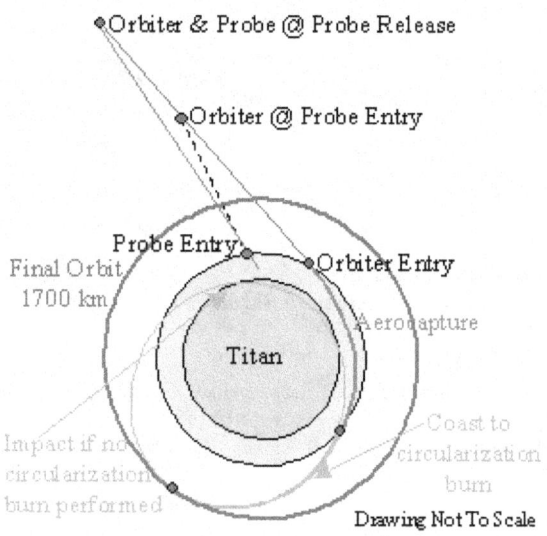

Figure 2.   Aerocapture Trajectory

The primary heat pulse of aerocapture lasts less than 10 minutes, during which the Orbiter is actively controlling its bank angle with hydrazine thrusters. After atmospheric exit the aeroshell is jettisoned and the Orbiter prepares for the periapsis raise maneuver to insert the Orbiter into a 1700 km circular orbit.

MISSION SYSTEM DESCRIPTION

The Orbiter Flight System is the primary focus of this study and the Ground Data and Mission Operations Systems were not addressed. The SEPM is largely inherited from previous study and the AP is treated as a black box. The Launch and SEP configurations are shown in Figures 3 and 4, respectively. The launch system mass summary is shown in Table 2. The post SEP cruise configuration is shown in Figure 5.

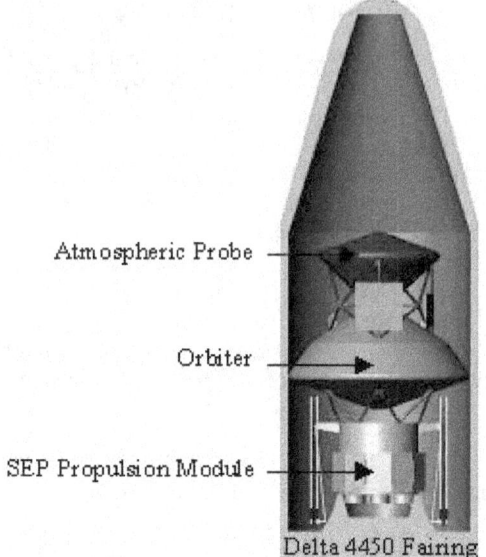

Figure 3.   Launch Configuration

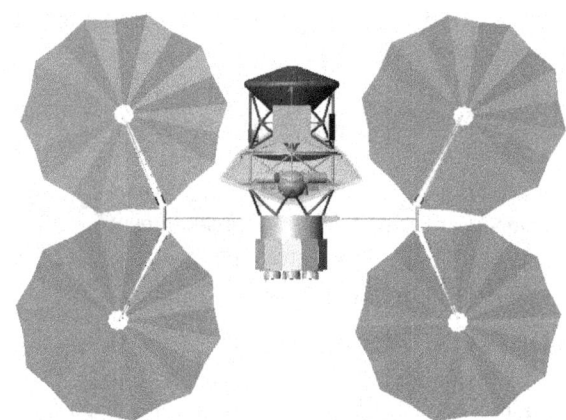

Figure 4.   SEP Burn Configuration

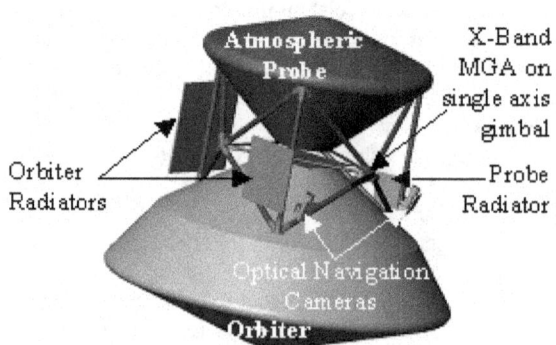

Figure 5.   Post SEP Cruise Configuration

Table 2. Launch Mass Summary (kg)

Element	CBE	GC	GE	A
Atmospheric Probe	280.2	29.8%	363.8	400.0
Orbiter/AP Interface	47.5	30.0%	61.8	61.8
Orbiter Dry Mass	743.0	28.4%	954.0	1200.0
Orbiter Prop Mass	140.6	2.2%	143.7	
SEP/Orbiter Interface	47.3	30.0%	61.4	61.4
SEP Dry Mass	623.9	29.9%	810.4	1450.0
SEP Prop Mass	460.1	10.0%	506.1	
Launch/SEP Interface	60.0	30.0%	78.0	78.0
Dry Mass (DM) Totals	1801.9	29.3%	2329.4	2601.4
Prop Mass (PM) Totals	600.7	8.2%	649.8	649.8
Stack Total	2402.6	24.0%	2979.2	3251.2
Launch Vehicle Capability	3423	LVC		
Launch Wet Mass Margin	29.8%	(LVC - CBE) / LVC		
System Reserve	13.0%	(LVC - Growth) / LVC		
Launch Dry Mass Margin	35.0%			

CBE = Current Best Estimate
GC = Growth Contingency = (GE - CBE) / CBE
GE = Growth Estimate
A = Allocation from system
Launch Dry Mass Margin = same as Launch Wet Margin with total propellant mass subtracted from all estimates.

Key Mission System Trades

Several trades associated with the overall Flight System are worth mentioning. These trades do not represent the a complete trade space for the Titan mission only those trades which drove the configuration of the flight system to allow convergence of a mission concept and determine aerocapture system feasibility.

Launch Stack Configuration

The orientation and placement of the AP and the Orbiter on top of the Propulsion Module (PM) drove the structural mass of the adapters as well as the primary structure mass for the PM and the orbiter. In general, 4 configurations were analyzed: the Orbiter and AP in nose up and nose down configurations with the Orbiter below and above the AP. The final configuration selected was Orbiter nose down (with respect to the launch vehicle) below the AP oriented nose up.

For all configurations of the Orbiter above the AP, the PM to Orbiter adapter became complex and massive. Additionally the large Orbiter mass suspended high on the launch stack resulted in much higher SEPM structure mass to accommodate the lateral launch loads and frequencies.

The Orbiter / AP orientation was selected as tail to tail for two reasons. First, there are potentially three separation planes between the Orbiter and the AP: Orbiter/AP, Orbiter/Truss, and AP/Truss. Each separation plane poses scarring risks to the TPS of the Orbiter and AP. Second, the structural interface of the truss with each vehicle is a risk to the TPS burn through (interface results in localized thermal anomalies). Routing the structural interface through the aft body lowers these risks.

The resulting stack configuration routes the primary PM to Orbiter adapter structure through the Orbiter fore body TPS. An engineering solution to the localized thermal anomalies at the interface with the TPS is considered solvable, but at the same time, it is considered highly desirable to find an alternate configuration which avoids perforation of all fore body TPS.

Cruise Propulsion System

A SEPM was selected over a chemical stage for the Earth to Saturn trajectory because of the SEP trajectory's superior overall performance in terms of delivered mass, flight time less than 6 years, and atmospheric entry velocity of around 6.5 km/sec. A detailed discussion of the chemical versus SEP trade is addressed by Noca, et al.[3]

AP / Orbiter Delivery

The delivery of the AP and the Orbiter to their respective entry trajectories could be performed by the Orbiter or by the SEPM. Since the aerocapture phase required the Orbiter to have all the subsystems required to perform AP and Orbiter delivery, the Orbiter was selected to perform these entry trajectory deliveries. This allowed deletion of the ACS, C&DH, and telecom subsystems from the SEPM and the separation of the SEPM soon after its burn out.

Probe Entry and Descent Data Relay

The AP critical event relay during entry and descent is a multi dimensional trade involving delivery errors (Orbiter and AP), telecom (AP and Earth), and Orbiter atmospheric entry risk (late separation of non entry system components). The selected strategy may not be the best solution, but is adequate to show feasibility for this study. The aspects of the Probe to Orbiter relay link are discussed in more detail in a later section.

Critical events relay using the SEPM on a flyby trajectory was ruled out because this would require the SEPM to be an independent spacecraft with unnecessary functional duplication with the Orbiter.

Titan Orbit Altitude

The initial desired science orbit altitude was specified at 1400 km. Orbit maintenance analysis performed by LaRC showed that for ballistic coefficients similar to the Orbiter design, as much as 100 m/s would be required for a three year mission – resulting in more propellant mass than the Orbiter could carry in our reference mission. The same LaRC analysis showed less than 2 m/s if the altitude was raised to 1700km. The only impact from a higher orbital altitude was to the science instruments. Since none of the specific instruments exist at this time, it was determined that the instrument impact was acceptable. Actual atmospheric density results from the Hyugens probe may provide an opportunity to lower this altitude if necessary.

ATMOSPHERIC PROBE DESIGN

As stated earlier, the AP design is considered to be a black box and out of the scope of this study. There are internal JPL studies performed by Team-X[12] indicating that 400kg is an adequate allocation for a Titan AP.

SEP MODULE DESIGN

A SEPM was selected over a chemical PM for the overall combination of shorter flight time, lower entry velocity, and lower PM mass; this trade is discussed in more detail by Noca[3]. An existing JPL SEPM design was modified for the Titan mission. The primary modifications were deleting the avionics in favor of using the Orbiter's avionics and increasing solar array structure and power capability. The SEPM structural mass was analyzed to assure proper launch load and frequency capability for the entire launch stack. The SEPM mass summary is shown in Table 3.

Table 3. SEP PM Mass Summary (kg)

Element	Flt Unit	CBE	GC	GE
SEP Wet Mass (kg)		1084.0	21%	1316.5
Xenon Propellant		460.10	10%	506.11
SEP Dry Mass	56	623.93	30%	810.40
ACS, C&DH, Telecom	3	5.85	30%	7.61
Power	7	148.31	30%	192.80
Structure	14	281.39	30%	365.81
Propulsion	20	142.28	30%	184.97
Thermal	12	46.10	28%	59.21

Most of the Attitude Control System (ACS), Command & Data Handling (C&DH), and Telecom functionality was moved to the Orbiter or to the Orbiter / Probe structural interface. This eliminates unnecessary component duplication through the entire system and allows the SEPM to be jettisoned after its job is complete. The majority of the Power system mass is the 25.6 m^2, 178 kg (growth), of solar arrays. The "Ftl Unit" column of Table 3 specifies the number of line items in the detailed mass list for the respective subsystem.

ORBITER DESIGN

The Orbiter design is shown in Figures 6, 7 and 8. Figure 6 shows the Orbiter in the Post AP release configuration. This configuration shows the critical components required by the Orbiter through post launch, SEP cruise, and AP entry and descent. These elements include Orbiter electronics and MMRTG radiators, X-Band MGA, AP UHF Relay antenna, and optical navigation cameras. Figure 7 shows the aerocapture configuration and Figure 8 specifies the primary Orbiter components.

Figure 6.   Post AP Release configuration

Figure 7.   Aerocapture Configuration

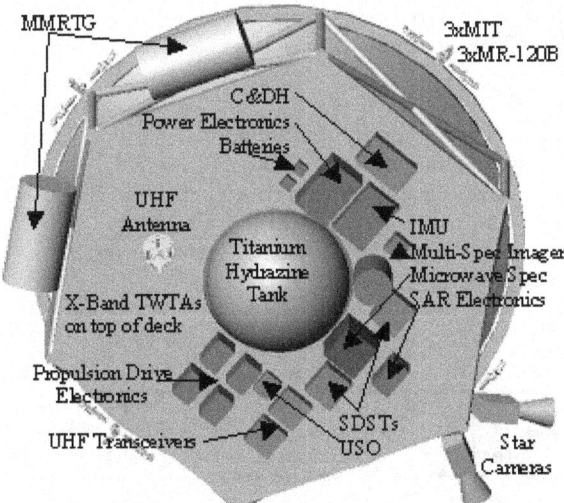

Figure 8. Titan Orbit Configuration

The estimated power required for the various mission phases is summarized in Table 4. Heater power in all phases is minimal because of an assumption that the MMRTG excess heat, ~3700W, can be distributed across the spacecraft well enough to not require the heater power typical for deep space missions. The modes listed in Table 5 are not all the modes identified in the study, just the ones that stress the system. The available power listed is the power output of two MMRTGs after 1.5% output degradation per year. It is assumed that once the telecom and instrument components have been turned on, that they are never turned completely off, but rather are placed in a low power standby mode when not in use. The 25% margin shown in the table is typically considered not viable for a pre-project, but specific opportunities to improve this are noted in the power subsystem section.

Table 4. Orbiter Average Power (W)

	Mission Phases				
	Cruise	Aero Capture	Lander Relay	Orbit Science	Earth Comm
Margin	38.1%	46.6%	32.4%	26.9%	25.0%
Available	226	226	222	214	214
Totals	140.00	120.60	150.00	156.50	160.50
Instrument	1.5	1.5	18.0	70.5	18.0
ACS	40.1	44.2	40.1	40.1	40.1
C&DS	14.9	14.9	14.9	14.9	14.9
Power	3	3	3	3	3
Propulsion	0	27	0	0	0
Telecom	76.5	30.0	66.0	20.0	76.5
Thermal	4.0	0.0	8.0	8.0	8.0

Table 5 presents the Orbiter mass summary with subtotals for Titan orbit, entry, and launch mass. All components are block redundant except for structure, propulsion, antennas, MMRTGs, thermal radiators, and science instruments. Generally, the Orbiter design was driven towards single fault tolerance without mission loss. Generally, a growth contingency (GC) of less than 30% in Table 5 indicates where components of high heritage are utilized in the system design. Instruments, structure, power, and thermal are the primary new development subsystems.

Table 5. Orbiter Mass List Summary (kg)

Orbiter Element	Flt Unit	CBE	GC	GE
Launch Wet Mass		883.6	24%	1097.7
Total Propellant		140.6	10%	143.7
Total Dry Mass	293	743.0	28%	954.0
Cruise Support Mass		65.5	9%	71.6
Propellant		41.4	0%	41.4
Dry Mass	49	24.1	25%	30.2
Structure & Misc	7	8.2	16%	9.5
Thermal	42	15.9	30%	20.7
Atmos Entry Wet Mass		818.1	25%	1026.1
Aerocapture System		416.7	24%	515.0
Propellant		88.9	0%	88.9
Bank Angle Control		12.6	0%	12.6
Circularization dV		76.3	0%	76.3
Dry Mass	8	327.8	30%	426.1
TPS	2	177.3	30%	230.5
Structure	6	150.5	30%	195.6
Titan Orbit Wet Mass		401.4	27%	511.1
Propellant		10.3	30%	13.4
Dry Mass	236	391.1	27%	497.7
Instruments	4	32.8	30%	42.6
ACS	15	20.2	10%	22.2
C&DH	16	15.3	26%	19.2
Power	5	80.1	30%	104.1
Telecom	13	39.1	24%	48.4
Structure	7	136.7	30%	177.7
Propulsion	73	39.5	21%	47.8
Thermal	103	27.4	30%	35.7

Key Flight System Trades

Reaction Wheels vs. Thrusters

The approach navigation and the science teams would prefer reaction wheels for attitude control rather than RCS jets. Unfortunately, the 2 MMRTG design could not supply enough power to operate 3 reaction wheels and everything else needed for science data gathering. Minimum Impulse Thrusters (MIT), TRL 6 in 2004, were selected as a compromise providing the necessary stability for science pointing, but degraded performance for approach navigation (small forces integration). The navigation analysis performed by Haw[4] considers a spacecraft with reaction wheels; this inconsistency was not resolved before the end of the study.

Rigid vs. Deployable X-Band HGA

The selected 2.4m Fixed HGA selected provides a 2.3 Tbit data return capability to a 70m station assuming one 8 hour contact every day. This falls well short of the ~11 Tbits generated by the science instruments. A 6 meter deployable HGA was investigated, but it was not able to be incorporated into the design before the end of the study. Follow-on work for a Neptune aerocapture mission has since determine a deployable HGA is a feasible concept for the Titan mission timeframe. A 6 meter antenna would provide a 500 kbps return link capable of returning 15 Tbits and provide an opportunity to reduce the aeroshell diameter and change the backshell design to a single angle. Such an antenna would increase pointing knowledge and control requirements, but these requirements are to be within the capability of the MIT RCS based ACS system. The larger antenna would also affect the SAR instrument design because of the narrower beam width – higher resolution, but possible less than global coverage.

Subsystem Descriptions

Orbiter subsystems will be discussed in order of overall system impact. In general, subsystems discussed first drive the system design more than those discussed last.

Aerocapture System

The aerocapture system is defined as the TPS, the underlying aeroshell structure, the propellant required for attitude control during aerocapture, and the propellant required for the orbit circularization burn at the apoapsis of the aerocapture exit orbit.

The aerocapture system structure, TPS and their associated aero-thermal design basis are described in more detail by Justus, et al[5] (Titan atmosphere), Masciarelli, et al[6] (Guidance algorithms), Way, et al[7] (Simulation), Takashima, et al[8] (Aerothermodynamics), Oleiniczak, et al[9] (Radiative heating), and Laub[10] (TPS).

The study team started with the largest possible aeroshell, 3.75m diameter, that would fit inside a 4m launch vehicle fairing in anticipation of needing a large diameter high gain antenna. As discussed earlier, changing to a deployable HGA would allow a smaller diameter aeroshell. The selection of the 70 degree cone angle is discussed by Masciarelli, et al[6].

Telecom

The Orbiter telecom system supports two primary links, X-Band to Earth and UHF to the AP. Table 6 summarizes the driving data return links. The X-Band System utilizes SDSTs and 50W TWTAs for communicating to Earth. The UHF telecom system is based on a next generation Electra Radio.

Table 6.   Orbiter Telecom Links

X-Band

Mission Phase	Orbiter Antenna (dBi)	Ground Station	Dist (AU)	Data Rate (bps)	Xmit Pwr (W)
Launch	-6, Patch	34m BWG	0.5	10	50
Cruise	24.8, Printed Dipole	70m	11	500	50
Aero capture	None				
Orbit Science	44.3, Fixed Dish	70m	11	75000	50

UHF

Mission Phase	Orbiter Antenna (dBi)	Probe Antenna (dBi)	Dist (km)	Data Rate (bps)	Xmit Pwr (W)
Probe Entry / Descent	-3 @ 30°, 2x2 patch	-3 @ 60°, Omni	85000	carrier only	4
Science Relay	5, Helix	-3 @ 60°, Omni	2050	64000	4

Power

At 10 AU, solar power was out of the question. A Mutli-Mission Radioisotopic Thermal Generator (MMRTG) unit was selected for the Orbiter power source. This unit is currently in development and should reach TRL 6 by 2006. The expected performance of the MMRTG is approximately 6.3% efficiency for a 2000W thermal input.

The efficiency of the units is tied to, among other things, the temperature differential of the unit. The 6.3% efficiency is related to a finned exterior radiating to space. Cooling the exterior of the unit below what is expected from the finned radiator design will yield better power output. There is a potential with the Orbiter's passive loop heat pipes to cool the surface of the MMRTGs to improve the power efficiency. To raise the power margin listed in Table 4 from 25% to 30% requires a power conversion efficiency of only 6.7%.

The power system includes 2 MMRTGs, total power available of 252W BOL, 214W at end of mission. Secondary batteries are included to help during peak periods with a typical assortment of battery charge controllers, power switching, and power conversion electronics.

ACS

Because of the limited power, reaction wheels were discarded in favor of thrusters capable of 0.7 mN-s impulses. All other ACS components are fairly standard equipment including star trackers, IMUs, and propulsion driver electronics. The Orbiter does not possess sun sensors because there is no critical need to sun point during a spacecraft upset. The Orbiter possesses 2 star trackers, 2 optical navigation cameras, and 2 C&DH strings which should suffice to allow the spacecraft to determine its attitude and point the HGA at Earth instead of solar panels at the sun during an off-nominal event.

Propulsion

The propulsion system is a blow down hydrazine monopropellant system with two sets of thrusters. The first set of thrusters is comprised of 12 MIT thrusters, each with a 0.7 mN-s minimum impulse capability to be used for fine attitude control. The MITs are currently in a flight qualification process on track for TRL 8 before 2006. The second set is comprised of 12 MR-120B engines, each with a 133.5 N force capability. The MR-120Bs are used for attitude control during aerocapture and for Titan orbit maneuvering.

Six thrusters are put on a dedicated line with latch valve, for a total of 4 latch valves, to ensure single fault tolerance (in degraded performance mode) against loss of mission. The single tank is a 74cm diaphragm tank with a Titanium shell.

C&DH

The C&DH system is JPL X2000[13] based. The cards selected are currently planned for TRL-6 by 2006, but an MRO based C&DH system might provide a lower risk technology solution in a sufficiently low mass and power package.

Thermal

The mission design presents several challenges for the thermal design:

1. The MMRTGs together generate over 3700W of thermal heat.
2. The MMRTGs are enclosed in an aeroshell designed to keep heat from getting in.
3. The radiator system has to be designed to work before, during, and after aerocapture
4. Inside the aeroshell, the system will experience solar distance of 0.95 AU (0.7 for Venus Gravity Assist) to 10 AU.

Because Venus gravity assists were considered in the mission trade space[3], a 0.7 AU minimum solar distance was assumed for the thermal design efforts. A ~30 node lumped mass model of the spacecraft was constructed to compute temperature distributions during the key mission phases for various design options. Titanium/Water loop heat pipes running to hot radiators mounted on the Orbiter / Probe truss were chosen to solve the problem of getting the heat out of the aeroshell, and these were found to work even in the 0.7 AU hot case at Venus. Aluminum / Ammonia loop heat pipes were also added to transport Orbiter electronics heat out of the aeroshell. A second set of Titanium/Water LHP carry MMRTG heat directly to the hydrazine tank.

The computational model results confirmed that all of the key avionics and propulsion components were maintained within prescribed operating temperatures during both the cruise to Saturn and after orbit insertion when the aeroshell was jettisoned and the orbiter exposed to the cold space environment at Titan.

For the aerocapture phase, it was assumed that the radiators were separated from the aeroshell 40 minutes before aerocapture and that the aerocapture lasted 20 minutes, a total of 60 minutes without radiators. The heat from the MMRTGs and from the high speed entry is simply absorbed by the thermal mass of the vehicle during this time. In the thermal analysis, the entry heating was approximated by an instantaneous jump to 250°C in the bondline temperature between the TPS and structure, which is a conservative assumption. The results demonstrated that all orbiter components were within their operational temperature ranges; although some with small margins.

Structure

The structure is discussed in more detail by Hrinda[11]. In general, the Orbiter primary structural design was driven by:

1. LV frequencies for Orbiter and Probe.
2. Combined geometry constraints of SEPM, Orbiter, and Probe in LV fairing.
3. LV loads for Orbiter and Probe.

NEW TECHNOLOGY DEVELOPMENT

Of all the technologies proposed that are currently less than TRL 6, only the TPS materials require additional funding to test the materials against the radiative heat loads expected at Titan[9]. The other technologies that are not currently at TRL 6: MMRTG, SEP Engine, and SEPM solar arrays, are all currently funded to reach TRL 6 in the 2006 time frame. If none of these three technologies actually reach TRL 6 by 2006, then the mission could still be performed with the currently lesser (mass) efficient technologies and a larger launch vehicle.

RECOMMENDED ADDITIONAL ANALYSIS

Many questions and trades consistent with continued Phase A/B efforts were identified by the study team. A summary of these issues is presented below along with a general classification of the issue as a lien, or opportunity, or either.

- **Launch Vehicle**: Verify 4m fairing not available for Delta 4450 (Opportunity).
- **Launch Configuration**: Eliminate structure through primary TPS (Lien).
- **SEP Propulsion Module**: 1) Incorporate latest Glenn Research SEP Engine capability (either). 2) Develop solar array deployment sequence concept and verify associated structures and mechanisms mass (Lien).
- **Atmospheric Probe**: 1) Verify 400kg is adequate (either). 2) Develop separation plane concept between AP and Orbiter which handles AP spin eject and thermal issues (AP MMRTG radiators) for 30 day coast to entry interface (Lien).
- **Navigation**: Verify use of MITs does not degrade navigation performance beyond mission requirements (Lien).
- **Science Instruments**: Develop conceptual designs for Multi-Spectral Imager and Microwave Spectrometer and verify TRL, mass, power, volume estimates (Lien).
- **Power**: 1) Develop detailed power modes and profiles (either). 2) Verify 2 MMRTGs are adequate for full mission (Lein). 3) Verify conversion efficiency of MMRTG based on thermal design (Opportunity). 3) Verify EMI/EMC compatibility for component configuration (either).
- **Thermal**: Verify MMRTG heat can be effectively routed to hydrazine system (manifolds, lines, thrusters) to eliminate need for heaters (Lien).
- **Telecom**: 1) Investigate trade between Ka-Band system or 6m deployable antenna for X-Band system (either). 2) Verify UHF line of sight for AP-Orbiter link during EDL and science relay are consistent with antennas and pointing concept (Lien). 3) Add LGA/MGA for Earth acquisition prior to high bandwidth links (Lein).
- **Aeroshell**: 1) Optimize packaging for smaller aeroshell (Opportunity). 2) Verify heating and TPS for new ballistic coefficient (either).
- **Cost**: Generate cost estimate for complete flight system (either).

CONCLUSIONS

The study demonstrates general technical feasibility for a Titan Explorer Orbiter flight system designed to use aerocapture as the Titan orbit insertion mechanism. Many liens exist against the conceptual design presented, but opportunities balancing the liens also exist. A change from the medium launch vehicle to a heavy lift launch vehicle would help retire many of the leins without invalidating the feasibility of the mission concept. Technology readiness for the flight system is good with all major components currently being funded to achieve TRL 6 by 2006 to support a possible launch date as early as 2010.

ACKNOWLEDGEMENTS

The spacecraft and mission design described herein were built upon a Titan aerocapture mission study performed by JPL's Team-X[12]. Although the configuration presented here is very different from the Team-X design, the Team-X design provided a valuable reference point from which system wide trades could be evaluated more easily.

Personnel from Langley Research Center, Johnson Space Center, Ames Research Center, Marshall Space Flight Center, and the Jet Propulsion Laboratory[3,4,5,6,7,8,9,10,11] were instrumental in determining system level requirements and subsystem capabilities for trade and mission performance analysis. Discipline experts at the Jet Propulsion Laboratory including Dave Hansen (Telecom), Brian Okerlund (Computer Aided Design), Jonathan Lam (Structural Analysis), John Huang (Antennas), Ray Baker (Propulsion), Rolando Jordan (SAR Instrument), Nick Mardesich (Solar Arrays), Bill Nesmith (MMRTGs); Applied Sciences Laboratory including Chern-Jiin Lee (thermal); Swales Aerospace including Michael Nikitnin (Loop Heat Pipes); all provided valuable input into the subsystem conceptual designs represented in this paper.

Funding for this work was provided by the NASA In-Space Propulsion Program, managed by Les Johnson, MSFC.

BIBLIOGRAPHY

1. Chyba, C. et al., "Report of the Prebiotic Chemistry in the Solar System CSWG", NASA internal report,1999.

2. Tamppari, L. et al., "Cassini/Huygens Follow-On

Mission Study Science Report", JPL internal report, 2001.

3. M. Noca, R.W. Bailey, C.J. Sauer, and R.E. Dyke, "Titan Explorer Mission Trades from the Perspective of Aerocapture", 39th AIAA/ASME/SAE/ASEE Joint Propulsion Conference, Paper Number AIAA-2003-4801, July 22, 2003.

4. R.J. Haw, "Titan Approach Navigation for the Titan Aerocapture Orbiter", 39th AIAA/ASME/SAE/ASEE Joint Propulsion Conference, Paper Number AIAA-2003-4802, July 22, 2003.

5. C. Justus, A. Duvall, D. Johnson, "Engineering Level Model Atmospheres for Titan and Neptune", 39th AIAA/ASME/SAE/ASEE Joint Propulsion Conference, Paper Number AIAA-2003-4803, July 22, 2003.

6. J. Masciarelli, E. Queen, "Guidance Algorithms for Aerocapture at Titan", 39th AIAA/ASME/SAE/ASEE Joint Propulsion Conference, Paper Number AIAA-2003-4804, July 22, 2003.

7. D. Way, R.W. Powell, J. Masciarelli, B. Starr, K.T. Edquist, "Aerocapture Simulation and Performance for the Titan Explorer Misson", 39th AIAA/ASME/SAE/ASEE Joint Propulsion Conference, Paper Number AIAA-2003-4951, July 22, 2003.

8. N. Takashima, B. Hollis, K. Sutton, M. Wright, J. Oleiniczak, "Preliminary Aerothermodynamics of Titan Aerocapture Aeroshell", 39th AIAA/ASME/SAE/ASEE Joint Propulsion Conference, Paper Number AIAA-2003-4952, July 22, 2003.

9. J. Oleiniczak, D. Prabhu, M. Wright, N. Takashima, B. Hollis, K. Sutton, "An Analysis of the Radiative Heating Environment for Aerocapture at Titan", 39th AIAA/ASME/SAE/ASEE Joint Propulsion Conference, Paper Number AIAA-2003-4953, July 22, 2003.

10. Laub, B., "Thermal Protection Concepts and Issues for Aerocapture at Titan", 39th AIAA/ASME/SAE/ASEE Joint Propulsion Conference, Paper Number AIAA-2003-4954, July 22, 2003.

11. G. Hrinda, "Structural Design of the Titan Aerocapture Mission", 39th AIAA/ASME/SAE/ASEE Joint Propulsion Conference, Paper Number AIAA-2003-4955, July 22, 2003.

12. T. Sweetser, et al., "Titan Lander Conservative Science 01-06", Team-X Report, June 4,5,8,15, 2001.

13. "Europa Orbiter/X2000 Avionics Development Industry Briefing", NASA/JPL, Pasadena, Ca, June 6, 2001.

TITAN EXPLORER MISSION TRADES FROM THE PERSPECTIVE OF AEROCAPTURE

Muriel Noca, Robert W. Bailey
Jet Propulsion Laboratory
California Institute of Technology
Pasadena, California
Contact: Muriel.noca@jpl.nasa.gov

A detailed Titan aerocapture systems analysis and spacecraft design study was performed as part of NASA's In-Space Propulsion Program. The primary objective was to engineer a point design based on blunt body aeroshell technology and quantitatively assess feasibility and performance. This paper reviews the launch vehicle, propulsion, and trajectory options to reach Titan in the 2010-2015 time frame using aerocapture and all-propulsive vehicles. It establishes the range of entry conditions that would be consistent with delivering a 360 kg entry vehicle plus a 580 kg orbiter to Titan. Results show that inertial entry velocities in the range of 5.3 to 6.6 km/s are to be expected for chemical and solar electric propulsion options with Venus and/or Earth gravity assists. Trip times range from approximately 6 years for aerocapture orbiters to 8-11 years for all-propulsive vehicles. In addition to trip time reduction, the use of aerocapture enables the mission with a Delta 4450 class launch vehicle as opposed to an all-propulsive orbit insertion approach, which requires a Delta IV heavy or Titan IV class launch vehicle.

INTRODUCTION

As part of the NASA In-Space Propulsion Program, aerocapture was investigated as an option for orbit insertion around Titan, the largest Moon of Saturn. This study involved several NASA centers and had for objective to conceptually design an aerocapture system for a generic orbiter/lander mission. This paper provides an overview of the mission trades performed during this study. The main objectives of the mission trades were to:

1. Identify potential mission architecture and trajectories for a launch circa 2010-2015, which meant to identify launch vehicle options, launch opportunities and sensitivities, and potential trajectories using chemical ballistic propulsion and solar electric propulsion (SEP);
2. Understand the sensitivities in flight time and Titan atmosphere's inertial entry velocities;
3. Provide a baseline trajectory and mission timeline.

The level of analysis for the mission trades varied from relatively very detailed, in the case of the aerocapture system and SEP trajectory optimization, to more parametric in the case of the chemical system. The approach was to survey as much as possible the trajectory trade space, both for chemical with multiple gravity assists and for SEP with a wide range of flight times and various gravity assist options. Once the trajectories were compiled, the delivered mass at Titan was calculated given the maximum performances of representative launch vehicles. This delivered mass was then compared to the actual mass needed for Titan orbiter and lander design, thus highlighting the benefits of aerocapture.

This paper first briefly describes the Titan Explorer lander and orbiter and then summarizes the transportation architectures considered. It then describes the findings for chemical and SEP transit to Saturn options, system and trajectories. It also describes briefly the aerocapture system and the baseline trajectory. Finally, it shows the overall architecture trade results.

SPACECRAFT DESCRIPTION

The science objectives and basic spacecraft concept of this Titan Explorer mission were based on previous studies performed internally at the Jet Propulsion Laboratory.[1] The mission includes a landed module and an orbiter.[2] The lander was considered here as a black box, and only the navigation aspects of carrying this lander were taken into account. The lander performs a direct entry, independently of the orbiter. The orbiter was designed to perform aerocapture and modified in the trades when a chemical insertion was performed instead. The baseline concept uses Solar Electric Propulsion (SEP) to reach Saturn/Titan. Figure 1 shows the launch configuration of the overall spacecraft.

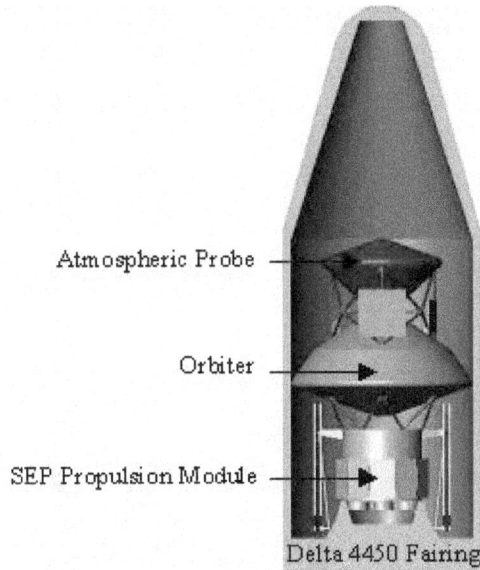

Figure 1: Titan Explorer spacecraft launch configuration.

TRANSPORTATION ARCHITECTURES

To understand the sensitivities in aerocapture entry conditions into Titan's atmosphere, it was necessary to perform a trade study of the various and most probable transportation options to Titan's orbit in the 2010-15 launch time frame. The transportation options for **launch and transit** from Earth to Titan were the following:

- Option 1: Ballistic with or without a chemical stage, launch to a high positive C3, with gravity assists.

- Option 2: Solar Electric Propulsion, launch to a low positive C3, with gravity assists.

The final science orbit around Titan was a 1700-km altitude circular orbit. Thus the **orbit insertion** options considered were:

- Option 1: Chemical insertion.

- Option 2: Aerocapture with a chemical burn for periapsis raise.

All four combinations of transportation were evaluated and will be described.

Aerobraking, which consists of low orbit insertion via several passes in a planetary atmosphere, was not considered at the time of the study. A more recent study on the possibility of performing aerobraking in Titan's atmosphere[3] has shown that aerobraking at Titan is limited by the gravitational perturbations of Saturn. The apoapsis of the aerobraking orbit would have to be quite low (below the 10000 km altitude range) to be in a gravitationally stable orbit around Titan. Higher apoapsis altitudes feature large spread in periapsis altitude (800 km spread in periapsis at an apoapsis altitude of 16000 km), making it very difficult to plan for and maintain aerobraking orbits. With this restriction, the delta-V saving of aerobraking compared to a direct insertion is quite low (~100-200 m/s), which limits its benefits.

CHEMICAL BALLISTIC TRAJECTORIES TO TITAN

Earth to Saturn Trajectories

Ballistic direct trajectories as well as gravity assist trajectories were computed and gathered by Jon Sims and Carl Sauer from the Jet Propulsion Laboratory Trajectory Group for a launch period between 2010 and 2016. Table 1 summarizes the performances of these trajectories. The maximum launch injected mass could then be found given the launch C3 for a Delta 4450 and a Delta IV Heavy. These launch vehicles were picked as representative of a range of launch vehicle performance. The launch vehicle data was provided by the NASA KSC Launch Support Group[4] and a 10% margin was held against the KSC provided performance (consistent with the JPL Team X conceptual design guidelines at the time of the study). Also note that some of the direct and gravity assist trajectories require a deep space maneuver, which Delta-V can sometimes be significant.

Table 1: Ballistic trajectories to Saturn/Titan.* EGA: Earth gravity assist, VEEGA: Venus Earth Earth gravity assist, VVVGA: Triple Venus gravity assist, JGA: Jupiter gravity assist, EJGA: Earth Jupiter gravity assist, VEEJGA: Venus Earth Earth Jupiter gravity assist.

	Launch date	Saturn Arrival Date	Flight time (yrs)	Vinf @ Saturn (km/s)	Approach declination (rad)	Vinf @ Titan (km/s)	Ventry inertial (km/s)	Deep Space DV (km/s)	Launch C3 (km2/s2)	Delta 4450 injected mass (kg) 10% margin	Delta IV Heavy included
Direct	2/4/16	11/29/20	4.8	6.26	-0.401	5.39	**5.84**	0	109.3	-	-
	1/16/14	2/14/20	6.1	5.25	-0.262	4.33	**4.88**	0	105.2	-	-
	1/7/12	1/22/20	8.04	5.48	-0.105	4.10	**4.67**	0	106.1	-	-
	1/3/11	12/20/19	8.96	5.82	0.017	4.23	**4.78**	0	107.2	-	-
	11/26/09	9/18/14	4.81	5.93	-0.105	4.36	**4.90**	0.513	108.7	-	-
	12/31/09	11/15/19	9.87	6.17	0.070	4.47	**5.00**	0	108.4	-	-
EGA	1/3/13	4/6/21	8.25	5.86	-0.349	4.96	**5.45**	0.407	48	-	3119
	3/3/13	4/27/20	7.15	7.3	-0.436	6.16	**6.55**	0.504	47.7	-	3141
	12/28/13	7/3/21	7.51	5.67	-0.332	4.80	**5.29**	1.488	28	2250	4905
	3/16/14	9/24/21	7.53	5.49	-0.140	4.16	**4.72**	1.693	26.3	-	5085
VEEGA	4/3/12	3/13/22	9.94	7.83	0.436	6.50	**6.88**	0	12.2	3150	6746
	4/3/12	3/13/22	9.94	6	0.436	5.39	**5.84**	0.217	12.1	3159	6759
VVVGA	7/26/10	12/3/18	8.4	9	-0.445	7.34	**7.67**	1.68	10.9	3249	6912
	8/1/10	11/19/19	9.3	6.83	-0.417	5.79	**6.21**	1.35	9.7	3339	7106
	7/1/13	12/31/23	10.5	6.5	-0.415	5.59	**6.02**	1.29	13.6	3051	6593
	6/1/15	5/27/25	10	6.36	-0.447	5.64	**6.07**	1.1	15.1	2943	6381
JGA	12/16/16	8/7/20	3.64	15.74	0.436	12.77	**12.97**	0	97.8	-	-
	12/16/16	8/6/22	5.64	7.28	0.436	6.15	**6.54**	0	82	-	-
	12/22/16	11/14/23	6.89	4.54	0.436	4.68	**5.19**	0	80.8	-	-
	12/23/16	7/19/28	11.57	2.48	0.436	3.98	**4.57**	0	76	-	-
EJGA	1/23/15	9/17/22	7.65	6.97	0.436	5.96	**6.36**	0.6695	25.6	2250	5162
VEEJGA	4/15/12	12/22/20	8.69	11.36	0.436	9.09	**9.36**	0	10.5	3272	6966

* The trajectories highlighted in yellow are the ones used in the subsequent analysis.

The trajectories highlighted in Table 1 were chosen for the overall mission trade as they represented a set of good performance trajectories. Careful consideration of all parameters (deep space Delta-V, approach velocity (Vinf) at Titan, launch C3, etc...) was used to make the choice of this set of trajectories.

The spread in entry velocity is depicted in Figure 2 for these trajectories. Thus for aerocapture purposes, most of the trajectories have inertial entry velocities between 5 and 7 km/s. This information will help select a baseline inertial entry velocity for a detailed aerocapture design, as will be discussed later.

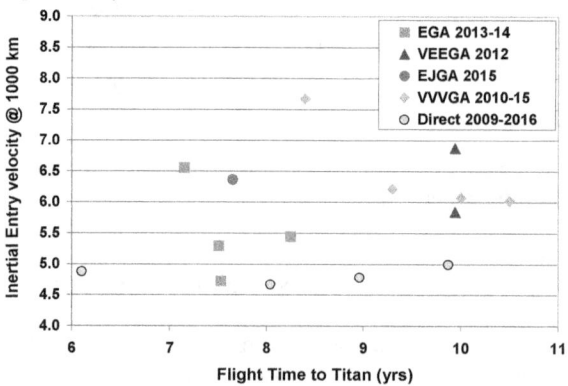

Figure 2: Titan inertial entry velocities for set of direct and gravity assist ballistic trajectories.

As can be seen from Figure 2, direct trajectories offer lower inertial entry velocities, since the time it takes to perform the gravity assist has to be made up for in the cruise to Saturn. Also, increasing the number of gravity assists (for instance from a single Earth gravity assist to a triple Venus gravity assist) increased flight time and launch mass, and if not, it then increased significantly the inertial entry velocity.

Chemical Insertion Delta-V assumptions

The chemical insertion Delta-V into a 1700-km altitude circular orbit around Titan can be computed given the Titan hyperbolic velocity (Vinf) and the approach declination. However, this Delta-V does not take into account a possible tour design around Saturn to pump down the initial orbit around Saturn using its satellites and to reduce the approach velocity around Titan. For the three trajectories picked, Table 2 summarizes the insertion Delta-V. The assumption was made that this Delta-V could be reduced by a moon tour down to 3 km/s with a flight time penalty of 1-2 years. No trajectory analysis has been done to verify this assumption, and further analysis needs to be done to confirm this estimate.

Table 2: Titan insertion ΔV for sample trajectories.

	Flight time (yrs)	Vinf at Titan (km/s)	Titan insertion ΔV (km/s)
EGA 2013	7.5	4.8	3.8
VEEGA 2012	9.9	5.4	4.3
VVVGA 2010	9.3	5.8	4.7

Chemical Propulsion system assumptions

To perform the chemical deep space maneuvers or insertion burns, a generic bi-propellant system was assumed. The dry mass for this system is summarized in Table 3. The specific impulse of the chemical system was assumed at 325 sec. In addition, 10% of the deterministic propellant mass was held as propellant contingency for maneuver clean-ups.

Table 3: Chemical propulsion system mass breakdown (includes 30% contingency).

Subsystem	Growth Mass (kg)
Not scaled with propellant mass:	
- Propulsion	19.5
- Thermal	16.5
- Telecom+electronics	2.3
- Structure	261.5
Scaled with propellant mass:	
- Tank	5%
- Tank structure	4%
- Thermal	1%

SEP TRAJECTORIES TO TITAN

Early on, an extensive database of Venus gravity assists SEP trajectories on a Delta 4240 was built, as they clearly provided better delivered mass for equivalent flight times compared to direct (no gravity assist) SEP trajectories. These trajectories served the purpose of evaluating the sensitivities in launch date, number of thrusters, power levels and inertial entry velocities. Subsequently, a smaller set of Earth gravity assist and Venus gravity assist on the Delta 4450 and Delta IV Heavy was calculated and used.

Carl Sauer (JPL) ran the SEP trajectory optimization code named SEPTOP for Solar Electric Propulsion Trajectory Optimization Program, which is based on the calculus of variations. This code optimizes two body interplanetary trajectories and can model discrete numbers of operating Xenon thrusters throughout the trajectory. Carl allowed for a coast time duty cycle of 10% to simulate times when the spacecraft is not thrusting due to housekeeping activities, and assumed a constant 250 W from the solar arrays for the spacecraft.

Solar Electric Propulsion system assumptions

The ion thruster used to calculate the SEP trajectories is and advanced 5-kW 5000 sec version of the flown NSTAR engine. The characteristics of the NSTAR technology can be found in many references.[5,6] The description of the 5-kW derivative of NSTAR named NGN for "Next Generation NSTAR" can be found in reference [7, 8]. This thruster is characterized by differences in four major parameters compared to NSTAR: engine input power, maximum specific impulse, and engine total impulse (or throughput) capability. Table 4 shows the projected performances of NGN.

Table 4: High-level NGN versus NSTAR characteristics.

	NSTAR	NGN
Max. thruster processed power (kW)	2.3	5
Engine diameter (cm)	30	30
Maximum Isp (sec)	3100	5000
Xe throughput (kg)	130	250

The ion propulsion system (IPS) was designed more as a propulsion module than just thrusters and power processing units. Figure 3 shows a simplified block diagram of the NSTAR IPS (single string). To that basic configuration was added redundancy, structural and thermal considerations. Figure 3 also shows an example of what the IPS module designed here could look like.

The solar arrays were sized based on a projection of the AEC-Able Ultraflex array capability. Since this array technology scales with power from ~ 1 kW up to ~ 30 kW, it was used as a representative potential technology for SEP applications. The specific mass was assumed to be 200 W/kg at 24 kW. A 14% degradation factor was applied to the array Beginning-of-Life (BOL) power to account for various degradation phenomena. Also, in order to support power demand during launch, a primary battery was used prior to solar array deployment.

The number of thrusters and PPUs was calculated on the basis of power requirements (4 minimum plus 1 redundant) and thruster propellant throughput. The system architecture followed a conventional approach with parallel strings of PPUs and thrusters. Each PPU drives one thruster but is cross-strapped to two engines. One spare ion engine, one spare PPU and DCIU were also included for single-fault tolerance. Each thruster was gimbaled separately. The PPUs were assumed to be 95% efficient.

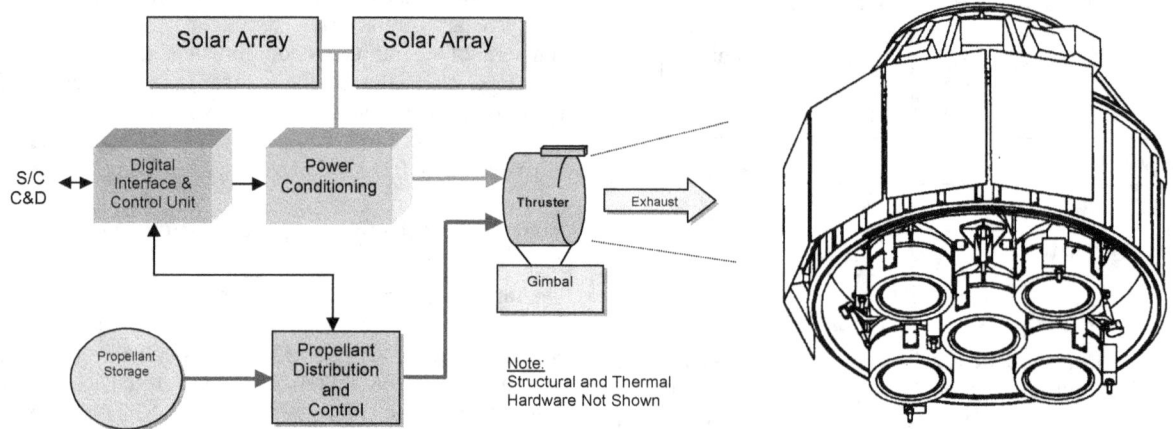

Figure 3: Ion propulsion module block diagram and conceptual configuration for system sizing.

The tank mass fraction was assumed to be 3.5% for Xenon when stored as a supercritical gas (~2000 psia). Furthermore, a 10% propellant contingency was added to the deterministic propellant mass to account for flow rate characterization, residuals, attitude control and margin.

Since the system masses are function of mainly power level, launch mass and propellant mass, each trajectory was uniquely considered and had a system mass associated with it. The component and subsystem sizing assumptions are given in Table 5. To be consistent with the JPL Team X conceptual design guidelines at the time of the study, 30% mass contingency was applied to all spacecraft subsystems, and a 10% launch vehicle margin was assumed.

Table 5: Ion propulsion system mass breakdown (includes 30% contingency).

Subsystem	Growth Mass (kg)
Not scaled with propellant mass:	
- Propulsion	168.5
- Power	192.8
- Thermal	59.2
- Telecom+ACS +electronics	7.6
- Structure	340.5
Scaled with propellant mass:	
- Tank	3.5%
- Tank structure	4%
- Thermal	1%

With the appropriate thruster model, trajectories were run for power level of 24 kW. Results are in terms of net delivered mass. The net delivered mass is defined as the spacecraft dry mass minus the dry mass of the ion propulsion system. Therefore the net delivered mass is everything on the spacecraft that isn't propellant or part of the ion propulsion module.

Net delivered mass and inertial entry velocity sensitivities to launch date, arrival date, SEP power level and thruster technology

Figure 4 shows results of net delivered mass as a function of flight time and launch years for the NGN thruster and for the NSTAR thruster (for reference). These trajectories were run for a Delta 4240 launch vehicle. The desired net delivered mass for the spacecraft was on the order of 2000 kg.

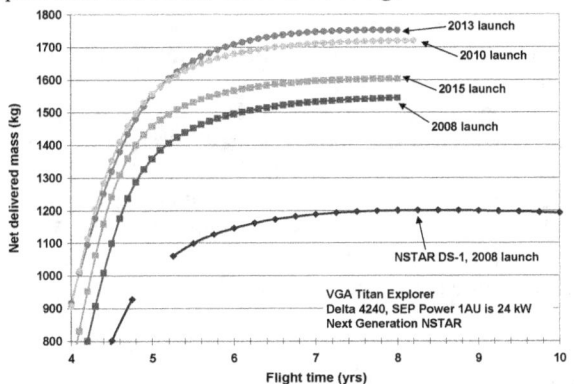

Figure 4: Venus Gravity Assist SEP trajectories to Saturn as a function of launch date.

Figure 5 shows the corresponding inertial entry velocities. As can be seen, the inertial entry velocity increases significantly for flight times below 5 years. It is also very dependent on flight time, launch date, and thruster technology. However, as Figure 6 shows, the inertial entry velocity was only weakly dependent on SEP power for a given launch date and thruster technology. There is also a significant variability in the choice of gravity assist. Thus choosing a flight time range will determine a range of inertial entry velocities. Flight times around 6 to 7 year offer the most "net delivered mass" benefit and result in entry velocities less than 7 km/s for most launch opportunities.

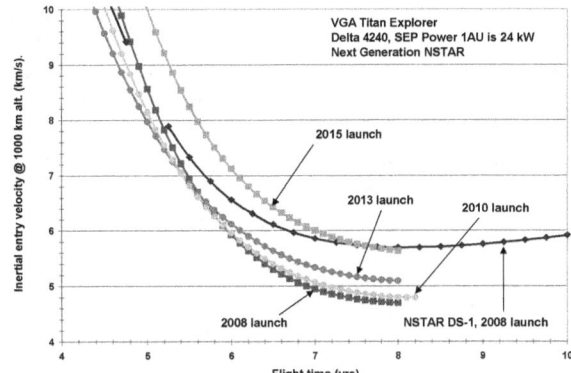

Figure 5: Titan inertial entry velocity at 1000 km altitude as a function of flight time.

The weak sensitivity to the SEP power level is mostly due to the fact that over the range of power looked at, the trajectory optimization code is going to try to follow the same optimum acceleration path. Thus for high power level, it will optimize the trajectory at lower launch C3, thus injecting more mass. The acceleration, which is proportional to the power level to mass ratio will be roughly the same as a low power, large C3, low launch mass case. Since it will follow almost the same trajectory profile, the arrival hyperbolic velocity will only vary slightly (such variation could be seen by zooming in Figure 6). This is true for a fixed flight time and launch date.

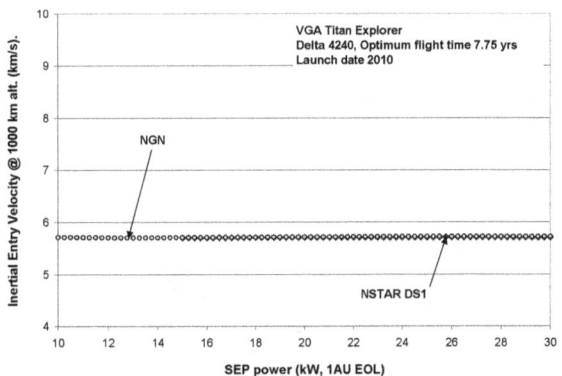

Figure 6: Inertial entry velocity at 1000 km altitude for Venus Gravity Assist SEP trajectories to Saturn with Delta 4240, fixed flight time (7.75 years) but varying SEP power level.

The launch window to perform a given gravity assist (VGA or EGA) is about one month. The sensitivity in propellant mass for that window is included in the 10% propellant margin.

Careful consideration was given to the sensitivity in arrival date. Since Titan's orbital period around Saturn is 16 days, the orbit geometry varies significantly depending on the arrival day. Figure 7 illustrates that point. However, it is possible to tune the arrival date with the SEP system or with a small chemical Delta-V at the end of the SEP phase to target a desired entry condition. The pattern shown in Figure 7 repeats every 16 days.

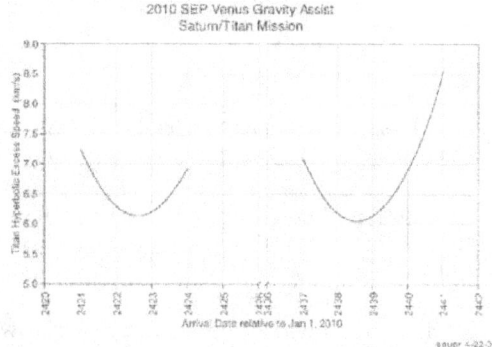

Figure 7: Sensitivity of Titan hyperbolic excess velocity as a function of arrival date.

Net delivered mass and inertial entry velocity sensitivities to gravity assist and launch vehicle

After looking at sensitivities in launch dates, arrival dates, SEP power level and thruster technology for a given gravity assist type and launch vehicle, the study called for more trajectories to perform the overall trade study. The SEP trajectories presented here assume a 2010 launch date and 24-kW SEP system with 4 operating NGN thrusters. The launch vehicle selected were the Delta 4450 and Delta IV Heavy, to enable more mass to be delivered, for both a Venus and Earth gravity assist. Figure 8 shows the net delivered mass for all 4 trade options. Figure 9 shows the corresponding inertial entry velocities. Also added to these figures are the points selected for the trade study.

Selection of the inertial entry velocity

In view of these results, it was decided that an entry velocity of 6.5 km/s would represent the best compromise between short flight times and high net delivered masses. Although somewhat arbitrary, it was felt that the aerocapture design would not change significantly for inertial entry velocities between 6 and 7 km/s. Using inertial entry velocities below 5.5 km/s or so would probably be feasible but would reduce the aerocapture performance and robustness.

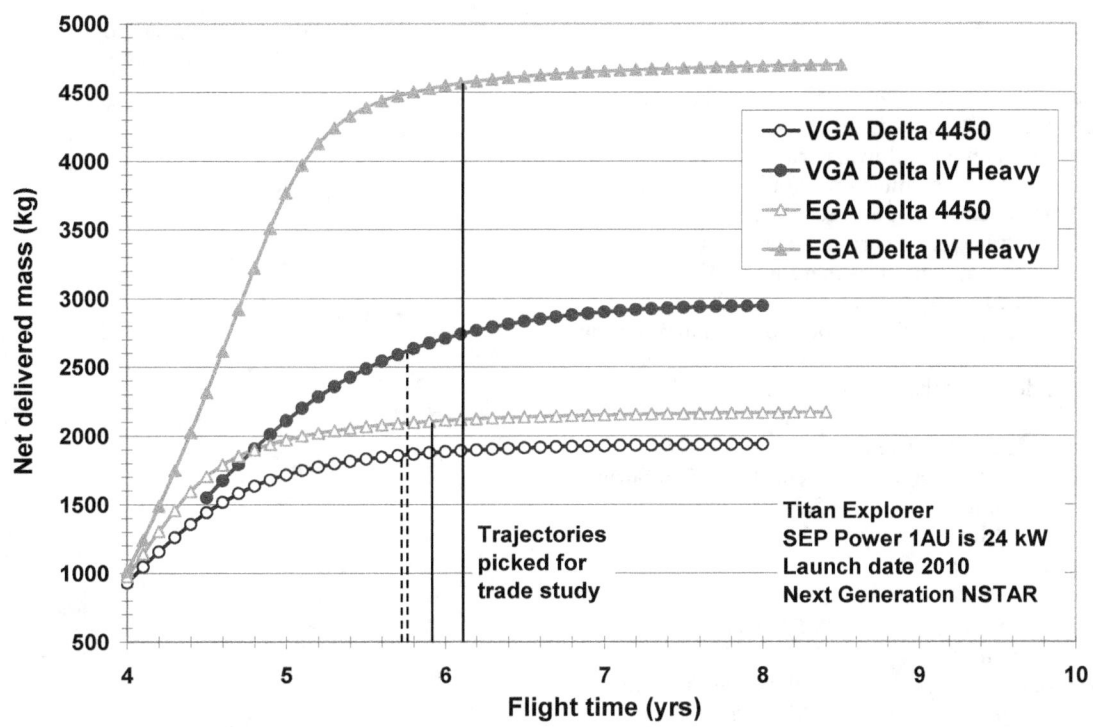

Figure 8: Venus and Earth gravity assist SEP trajectories to Saturn as a function of launch vehicle.

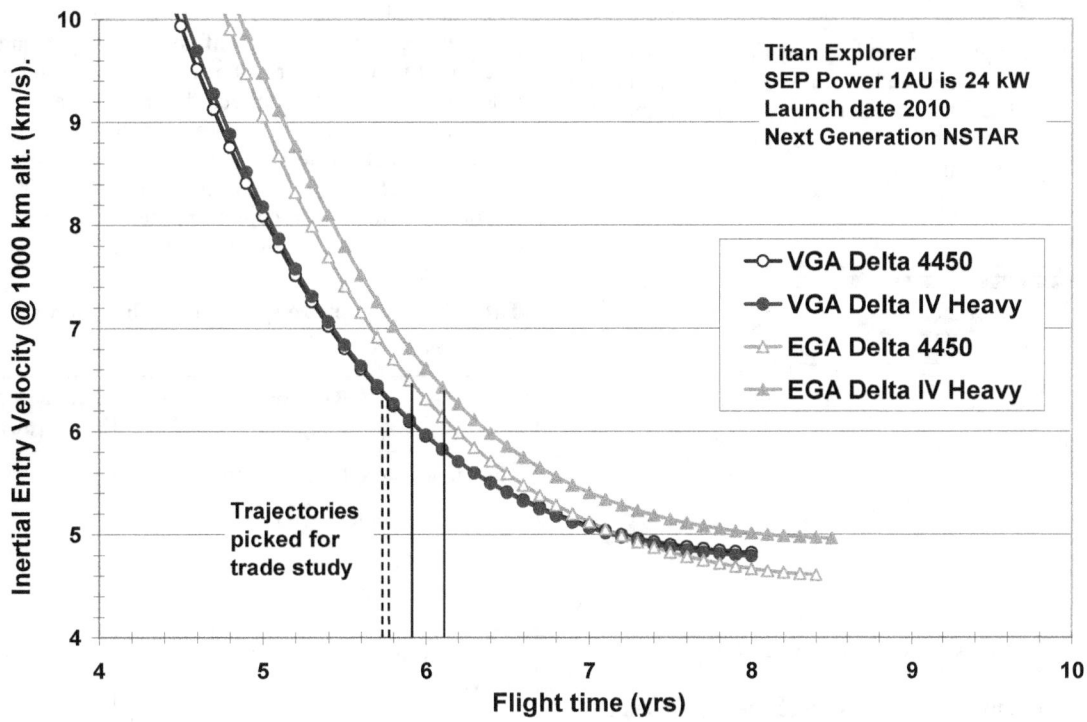

Figure 9: Titan inertial entry velocity at 1000 km altitude as a function of flight time.

Baseline trajectory

The baseline trajectory for the design of the aeroshell and other components of the aerocapture system was selected based on the following criteria:
- The mission architecture should use the smallest launch vehicle possible to reduce cost;
- The trajectory performance should provide adequate system mass margin (30%) for growth, and adequate system reserves (> 10%);
- The trajectory should provide a Titan inertial entry velocity close to 6.5 km/s consistent with the design of the aeroshell.

Thus the trajectory selected featured (see Figure xx):
- Launch vehicle: Delta 4450 (5 m fairing)
- Flight time: 5.9 years
- Launch date: 12/24/2010
- Arrival date: 11/17/2016
- Earth Gravity Assist: 12/03/2012
- Launch C3:   8.6 km^2/s^2
- Launch mass: 3423 kg
- Propellant mass: 460 kg deterministic
- Vhyp @ Saturn: 7.97 km/s
- V$_{entry}$ inertial: 6.5 km/s @ 1000 km
- Thrusters: 4 operating NGN
- SEP power level: 24 kW

A detailed approach and navigation analysis was performed[9] and suggested the timeline shown in Figure 11 for this baseline trajectory. This timeline includes Trajectory Correction Maneuvers (TCM) for both the lander and the orbiter targeting.

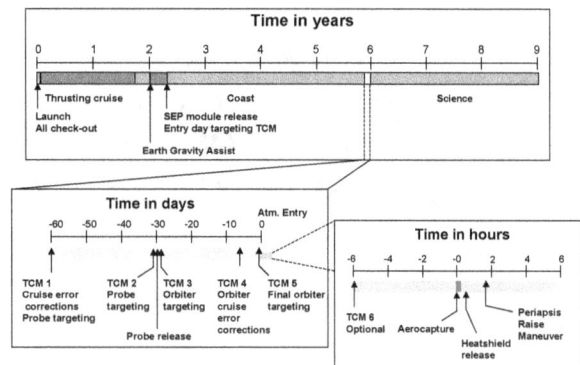

Figure 11: Baseline trajectory timeline.

AEROCAPTURE SYSTEM

The aerocapture system is described in detail in reference [2] and [10]. It has been designed for the baseline trajectory, which lead to an inertial entry velocity of 6.5 km/s.

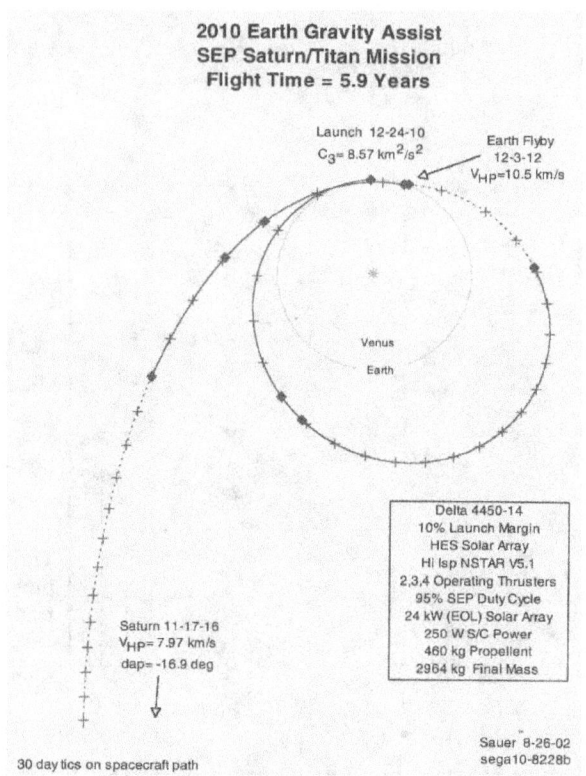

Figure 10: Baseline SEP trajectories to Saturn.

The heatshield design was based on a 70 deg. half cone-angle using the Viking-Pathfinder heritage, and was sized to fit the orbiter 2.4 meter diameter high gain antenna.. Table 6 summarizes the mass breakdown for the aeroshell system. As can be seen, the total dry mass of the aeroshell system is 426 kg for a total entry dry mass of 1026 kg (~ 41.5% aeroshell entry dry mass fraction). Other aerocapture-related hardware was ejected before entry. This hardware is also summarized in Table 6.

Table 6: Aerocapture system mass breakdown (includes 30% contingency).

Subsystem	Growth Mass (kg)
Mass that entered the atmosphere:	
- Heatshield, backshell and structure	426
- Hydrazine propellant	89
Aerocapture mass jettisoned prior entry:	
- ACS, telecom, thermal radiators and loop heat pipes for the RTG, intruments	72

MISSION ARCHITECTURE TRADE RESULTS

The overall mission architecture trade results are summarized in Table 7. This table shows first the type of launch vehicle followed by the gravity assist type, the transit propulsion system and the Titan capture system. It assumes that the full capability of the launch vehicle is used and calculates the payload surplus or deficit mass compared to the mass required at Saturn before insertion.

Table 7 also shows additional structure mass not part of the aerocapture system. The Pre-insertion ejected mass includes about 62 kg for the orbiter to lander interface structure and about 71 kg for ACS, telecom, thermal radiators and loop heat pipes for the MMR RTG and other. In the case of chemical insertion, the orbiter to lander interface is assumed not to be jettisoned and thus is included in the Payload in Titan orbit mass. The detailed mass breakdown can be found in [2]. Figures 12 and 13 render some of the results of Table 7. They show the **payload surplus or deficit mass** as a function of transit propulsion, gravity assist and launch vehicle for chemical or aerocapture insertion. The payload surplus or deficit mass is the mass above or below the necessary mass to deliver the lander and orbiter around Titan. It does not include the lander or orbiter mass. Both figures clearly show the advantages of aerocapture, which in every case looked at provided more payload reserve and shorter flight times than for a chemical insertion burn. However, they also show that it is possible to deliver sufficient payload mass (low margin) with an all chemical insertion system. Here again, the penalty will be flight time.

CONCLUSIONS

This paper summarizes the transit trajectory options for the Titan Explorer and derives the range of entry conditions for the aerocapture maneuver inside Titan's atmosphere. This survey shows that inertial entry velocities in the range of 5.5–7 km/s are to be expected. This range offers the best combination of highest delivered mass to Titan's orbit and lowest entry heating. The study chose to baseline an inertial entry velocity of 6.5 km/s for the detailed design of the aerocapture system, and the corresponding SEP trajectory is provided.

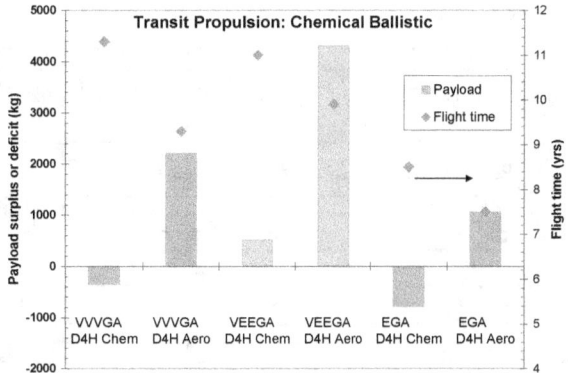

Figure 12: Payload surplus or deficit mass for the chemical ballistic transit cases for the Delta IV Heavy.

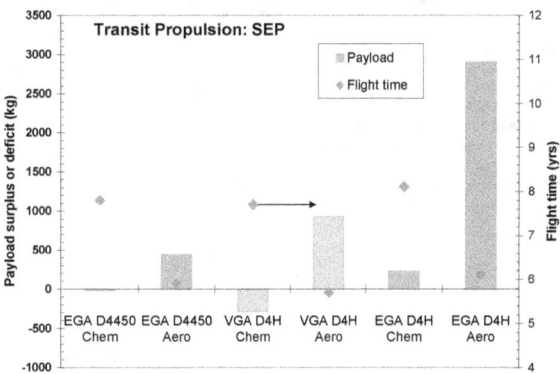

Figure 13: Payload surplus or deficit mass for the SEP transit cases for the Delta 4450 and Delta IV Heavy.

This paper also summarizes the mission transportation trades performed during the study to show the benefits of aerocapture. The study shows that aerocapture as a norbit insertion option provides more delivered mass in every launch vehicle and gravity assist case looked at than chemical insertion, and shorter flight time (typically by 2 years). However, all chemical or SEP with chemical insertion cases exist that would deliver the necessary mass in Titan's orbit with a Delta IV Heavy with flight times between 8–11 years. The baseline trajectory case for this study is an SEP aerocapture case on a Delta IV 4450 with a flight time of 5.9 years.

Table 7: Overall mission architecture trade results.

Launch Vehicle	Delta 4450					Delta IV Heavy								
Gravity Assist	VVVGA	VGA	EGA			VVVGA	VEEGA	VEEGA	VGA		EGA			
Earth to Saturn Prop System	Chem	SEP	Chem	SEP	SEP	Chem	Chem	Chem	SEP		Chem		SEP	
Titan Capture	Aero	Aero	Aero	Chem	Aero	Chem	Aero	Chem	Chem	Aero	Chem	Aero	Chem	Aero
Cruise Time to Titan (yrs)	9.3	5.7	7.5	~7.8	5.9	~11.3	9.3	9.9	~11	~7.7	5.7	~8.5	7.5	~8.1	6.1
Launch C3 (km2/sec2)	9.7	10.2	28	8.4	8.6	9.7	9.7	12.1	12.1	36.2	36.2	28	28	17.8	17.8
SEP Power (kW)	---	24	---	24	24	---	---	---	---	24	24	---	24	---	24
Entry Velocity (km/s)	6.2	6.4	5.3	5.3	6.5	6.2	6.2	5.9	5.9	6.4	6.4	5.3	5.3	6.5	6.4
Earth to Saturn Cruise ΔV (km/s)	1.35	8.3	1.49	6.4	6.7	1.35	1.35	0.2	0.2	6.9	6.9	1.49	1.49	4.6	4.6
Titan Insertion Chem ΔV (km/s)	0	0	0	3	0	3	0	4.3	3	3	0	3	0	3	0
Launch Capability	3339	3298	2250	3439	3423	7106	7106	6759	6759	4135	4135	3141	5450	6019	6019
Propellant Mass[1]	1239	619	893	1485	506	5464	2671	5160	4363	2015	715	2253	2208	3257	633
LV to Prop Module Adapter	78	78	78	78	78	78	78	78	78	78	78	78	78	78	78
Prop Module Dry Mass	424	820	389	808	810	846	567	816	736	1257	828	525	521	1383	821
Available Payload Mass to Saturn	1598	1781	890	1068	2029	718	3790	705	1582	784	2514	285	2644	1301	4488
Prop Mod to Payload Adapter	61	61	61	61	61	61	61	61	61	61	61	61	61	61	61
Direct Entry Payload Mass (Lander)	364	364	364	364	364	364	364	364	364	364	364	364	364	364	364
Payload in Titan Orbit (Orbiter)	511	511	511	573	511	573	511	573	573	573	511	573	511	573	511
Aerocapture System[2,4]	515	515	515		515		515				515		515		515
Pre-Insertion Ejected Mass	133	133	133	73	133	73	133	125	73	73	133	73	133	73	133
Required Mass to Saturn	1584	1584	1584	1071	1584	1071	1584	1123	1071	1071	1584	1071	1584	1071	1584
Payload Surplus (Deficit) Mass	14	197	(694)	(3)	445	(353)	2206	(418)	511	(287)	930	(786)	1060	230	2904
System Mass Margin					29.8%	(LV - CBE) / LV									
System Reserve					13%	(LV - Growth) / LV									

Assumptions and Notes:
All masses are growth mass listed in kg
[1] Propellant mass calculated using "Launch Capability" as system total mass; Isp = 325, includes 10% mass contingency
[2] Aerocapture mass for Chemical Earth to Titan Prop Modules may change slightly (entry velocity not equal to 6.5 km/s)
[3] This launch capability is extrapolated data
[4] Aerocapture mass includes propellant for circularization delta-V
[5] Assumes delta V required for capture can be reduced to 3 km/s through Saturn/moon tour with flight time penalty. No supporting analysis
[6] Propellant mass and Prop Module Dry Mass for SEP / Chem options includes propellant and dry mass for both SEP and chemical stages
[7] Titan Aerocapture Study Reference Mission

ACKNOWLEDGMENTS

The research described in this paper was carried out at the Jet Propulsion Laboratory, California Institute of Technology, under a contract with the National Aeronautics and Space Administration.

The authors gratefully acknowledge the contributions of Carl Sauer, Jon Sims, and Theresa Debban for all their trajectory creativity and calculations. The authors also wish to thank Eric Dyke (Swales Aerospace) and Mary-Kae Lockwood (LaRC) for their conscientious review of the results, John Brophy (JPL) for all the SEP thruster details, Ray Baker (JPL) for all the chemical system details, and the In-Space Propulsion Program for making the funding available for this work.

REFERENCES

[1] T. Sweetser et al., "Titan Lander Conservative Science 01-06", Team X report, June 4-15, 2001.
[2] R. Bailey, J. Hall, T. Spilker, "Titan Aerocapture Mission and Spacecraft Design Overview", AIAA-2003-4800, 39th AIAA/ASME/SAE/ASEE Joint Propulsion Conference, Huntsville, AL, July 2003.
[3] Communication with Greg Whiffen, JPL, during the In-Space Propulsion SAIC Titan Explorer review, April 2003.
[4] http://elvperf.ksc.nasa.gov/elvMap/
[5] Rayman, M. D., "Results from the Deep Space 1 Technology Validation Mission," IAA-99-IAA.11.2.01, 50th International Astronautical Congress, 4-8 Oct 1999, Amsterdam, The Netherlands.
[6] Polk, J. E., et al., "In-Flight Performance of the NSTAR Ion Thruster Technology on the Deep Space One Mission," AIAA-99-2274, 35th AIAA/ASME/SAE/ASII Joint Propulsion Conference, June 21-23, 1999.
[7] Brophy, J.R, "Ion Propulsion System Design for the Comet Nucleus Sample Return Mission," AIAA-2000-3414, 36th AIAA/ASME/SAE/ASII Joint Propulsion Conference, July 2000.
[8] M Noca, "Next Generation Ion Engines: Mission Performances", IEPC-03-313, 28th International Electric Propulsion Conference, Toulouse, France, March 17-21, 2003.
[9] R. Haw, "Approach Navigation for a titan Aerocapture Orbiter", AIAA-2003-4802, 39th AIAA/ASME/SAE/ASEE Joint Propulsion Conference, Huntsville, AL, July 2003.
[10] M. Lockwood, "Titan Aerocapture Systems Analysis", AIAA-2003-4799, 39th AIAA/ASME/SAE/ASEE Joint Propulsion Conference, Huntsville, AL, July 2003.

APPROACH NAVIGATION FOR A TITAN AEROCAPTURE ORBITER

Robert J. Haw
AIAA Member
Navigation and Mission Design
Jet Propulsion Laboratory, California Institute of Technology
Pasadena, CA 91109-8099
robert.haw@jpl.nasa.gov

ABSTRACT

A proposed mission will send an orbiter and surface probe to Titan. Aerocapture technology will slow the spacecraft at Titan and perform the orbit insertion. The navigation strategy uses ΔVLBI measurements and optical imaging in order to satisfy the flight system performance requirements. The performance requirement metric is the spacecraft's atmosphere entry flight path angle, which must fall within -36.8° ±1.0 (3σ) (the error bar is preliminary). The requirement can be satisfied with a data cutoff 7 days before Titan, assuming a Cassini-era Titan ephemeris. There is margin in the arrival template to tighten (reduce) the entry corridor by scheduling a data cutoff closer to Titan, and a data cutoff 2 days before Titan is baselined here. Improvements to the performance are discussed by anticipating enhancements to the current level of technological readiness. The surface probe can satisfy an entry flight path angle requirement of -50°±5 (3σ) by separating any time within ~3 months of Titan.

INTRODUCTION

The Titan orbiter mission will utilize solar electric propulsion during interplanetary cruise to deliver an orbiter and probe to Titan, and will employ aerocapture technology to assist with orbit insertion. The work presented here is part of a Titan mission study described in an overview paper given in Reference 1.

The focus of this paper is a demonstration of the feasibility of direct-entry aerocapture as a replacement for an orbit insertion maneuver. The probe is considered secondarily only in so far as illustrating that probe requirements can be met.

This paper examines the navigation accuracies of the orbiter and probe as they approach and encounter Titan. Since there is a limit on the accuracy with which the initial state and subsequent dynamics are known, the future state cannot be computed with certainty from the initial one. The 'delivery' is defined as the uncertainty expected in the future spacecraft state (at its time-of-arrival) computed at time T (where T is before the time-of-arrival). That is, a Titan–2 day delivery represents the prediction (with dispersions) of the location of the spacecraft at arrival, when 2 days away from Titan.

The epoch for this analysis is Titan-75 days.

SPACECRAFT CONFIGURATION

The design of the flight system is discussed in Reference 2. Externally, the flight system during Titan approach consists of two stacked aeroshells connected via an external truss. (The SEP propulsion module and associated solar array fans have been jettisoned earlier.) The spacecraft configuration is shown in Figure 1. The probe aeroshell is located on top and the orbiter aeroshell beneath. The probe and orbiter respectively are enclosed by the aeroshells, which consist of a backshell/ heatshield pairing. The aeroshell protects the contents from the high heat loads experienced during atmospheric entry.

An external truss connects the two aeroshells. Mounted on the truss is a rear-looking medium gain antenna (MGA) for telecommunications and navigation. Also mounted on the truss are forward-looking cameras for optical navigation. The cameras should have independent pointing control.

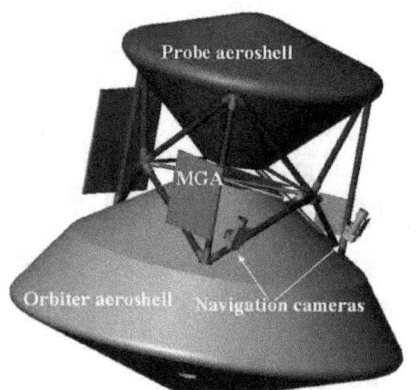

Figure 1. Spacecraft Configuration

The mass of the flight system (including propellant load) is 1465 kg. The orbiter represents 75% of the mass allocation and the probe/truss the remaining 25%.

ATTITUDE MAINTENANCE

The spacecraft is assumed to be 3-axis stabilized with a momentum-wheel ACS. The momentum wheels maintain spacecraft pointing. Balanced 0.7 N minimum impulse thrusters perform scheduled momentum de-saturation burns. (Note: momentum-wheel stabilization is not consistent with the baseline spacecraft design in Reference 2, which assumes limit-cycling ACS. The inconsistency has been identified but insufficient time was available to address this issue before the study ended.)

SEP MODULE

The SEP module consists of five Glenn Research Center 5 kW ion engines. Four engines operate at a time, leaving one engine in reserve. Each engine produces 0.15 N of thrust. Four solar array fans with a total area of 23 m^2 can generate 24 kWe at 1 a.u. [3]. The engines thrust while within the inner solar system, but when the solar range exceeds ~4 a.u. the SEP stage (including solar arrays) is discarded.

PROPULSION

The propulsion system after the SEP is jettisoned is a monopropellant hydrazine system. This subsystem must perform spin control, attitude control, and trajectory correction maneuvers (TCMs) during the approach to Titan.

Thrusters are used to de-saturate momentum wheels and to correct trajectory errors. All propulsive functions are performed with the RCS thrusters acting through ports in the backshell of the orbiter aeroshell (ports are not shown in Figure 1). After orbit insertion (without the aeroshell) the RCS thrusters perform the Titan pericenter raise maneuver.

TELECOMMUNICATIONS

The telecommunications subsystem operates an X-band system for direct-to-Earth communications and a UHF system for communications with the probe during entry, descent and deployment (EDD).

A truss-mounted X-band medium-gain antenna (MGA) (with a low gain antenna back-up) is the only communication link during interplanetary flight. The MGA design is a 24 dBi phase array antenna 28 cm in diameter. At 10 a.u. from Earth the data rate to a 70 m ground antenna is 600 bps [2]. The probe backshell includes an aft-mounted LGA for UHF communications during EDD.

Tracking and telemetry will use the Small Deep Space Transponder, which supports phase coherent two-way doppler and ranging, command signal demodulation and detection, telemetry coding and modulation, and differential one-way range (DOR) tone generation (tone sidebands at ±19 MHz). DOR tones are used for interferometric ΔDOR measurements.

TRAJECTORY DESIGN

The interplanetary reference trajectory characteristics are described in Reference 2. Since the reference trajectory has some fluidity during advanced study exercises (and changes frequently), the launch date chosen for this navigation investigation is December 15, 2010, and may not agree with the latest reference trajectory. The launch mass is 2515 kg, including the SEP module. The interplanetary trajectory includes a gravity assist flyby of Venus in May 2012 and arrives at Titan on August 21, 2016. The hyperbolic excess velocity at arrival is 6.1 km/s and the range to Earth is 9.8 a.u. (one-way light time equals 81 minutes).

The orbiter and probe initially approach Titan on an impact trajectory. After the probe separates from the orbiter bus it follows a ballistic path into Titan. The probe is nominally targeted to a mid-latitude region in the northern hemisphere (target TBD).

Before probe release the spacecraft is spun-up to a rate TBD (on the order of 3 rpm). After probe release the remaining orbiter bus and truss are spun-down and a trajectory deflection maneuver (TDM) is performed. See Figure 2, where on the scale of that figure the probe release and TDM appear to occur simultaneously. The truss is jettisoned before aerocapture.

The TDM serves two purposes: re-directing the orbiter's trajectory to intercept the orbiter entry interface point (a lateral movement of approximately 600 km in the B-plane), and slowing the orbiter with respect to the probe (a delay on the order of hours). This delay permits the orbiter to function as a communication relay for the probe during EDD.

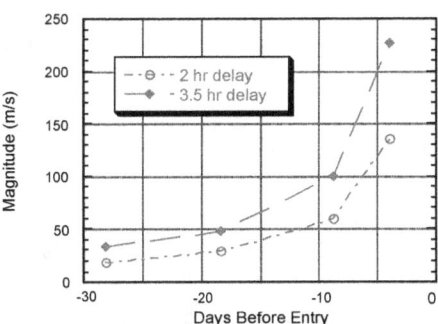

Figure 3: TDM magnitude v. probe release time

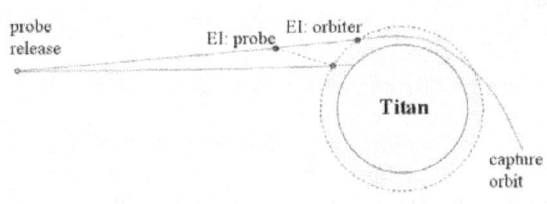

Figure 2: Entry Interface for Probe and Orbiter

Orbiter delay is a trade-off between probe transmitter power and the duration of EDD and initial surface operations (to be relayed and recorded by the orbiter before aerocapture). In Figure 2 the blue line indicates the orbiter and the red line the entry probe. The position of the orbiter at the time of probe entry is indicated by the blue dot with the label "EI: probe" (Entry Interface for probe). For a delay of 3.5 hours, when the probe reaches its entry interface, the distance between probe and orbiter is approximately 80,000 km. The orbiter subsequently enters Titan's atmosphere at the blue dot labeled "EI: orbiter". By this time the probe has been transmitting data to the orbiter for about 3.5 hours.

The magnitude of the TDM as a function of probe release time and slow-down maneuver is parameterized in Figure 3. A probe release at E-30 days and TDM at E-29 days represents the baseline unless otherwise stated.

The encounter with Titan is a direct-entry arrival, phased to arrive at the satellite when Titan is receding from the spacecraft along the spacecraft's approach asymptote (*i.e.* at a point in Titan's orbit near the minimum spacecraft-Titan relative velocity). The arrival at Titan is indeed direct – there is no Saturn orbit insertion occurring prior to TOI (*e.g.* as will happen with Cassini). Titan's north pole is targeted in order to place the spacecraft into a polar orbit. The orbiter's closest approach to Saturn occurs 13 hours before Titan-arrival at a range of 1.2×10^6 km (approximately equal to Titan's orbital radius).

The trajectory during approach is ballistic and the spacecraft depends upon chemical thrusters for flight path control.

TARGET DESIGN

The entry interface target consists of three parameters: inertial entry flight path angle (FPA), B-plane angle, and entry interface radius, where the interface entry radius is defined to equal 3575.0 km *i.e.* an altitude of 1000 km. The flight path angle is the angle subtended by the vehicle trajectory with the local horizontal at the EI, and it defines a corridor through the atmosphere (see Figure 4). The corridor's width is constrained by upper and lower bounds determined by the physical limitations of the flight system (vehicle must be able to withstand aerodynamic, structural, and heat loads), and by the need to accumulate sufficient drag forces to slow the spacecraft (to avoid skip-out).

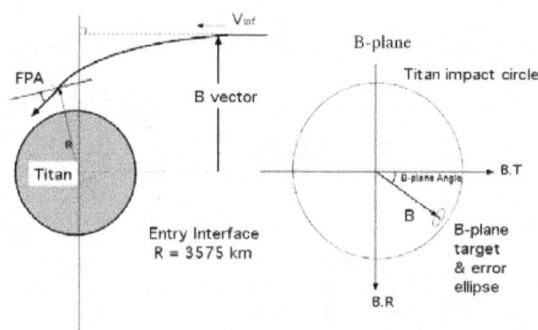

Figure 4: FPA v. B-plane

For the orbiter, the entry interface target is derived from the characteristics of the vehicle aerodynamics and the post-insertion Titan orbit. For the probe, the entry interface target is derived from the desired landing site (latitude and longitude).

The entry flight system at Titan is designed for an inertial entry velocity of 6.5 km/s and an entry FPA of -36.8°± ~1.0 (3σ) [4]. The error bars on the FPA are preliminary. The probe is designed to satisfy an entry FPA of -50°±5 (3σ) [4].

The atmosphere at Titan (mostly nitrogen and methane) has a surface density of 5 kg/m^3 (5x Earth), corresponding to a surface pressure of ~1.5 bar [5]. Subject to these conditions, an FPA of -50° subjects the probe to a maximum deceleration of ~10 g's. An FPA of -36.8° subjects the orbiter to a maximum deceleration of ~4 g's (assuming the baseline atmospheric density profile described in Reference 6).

The orbiter descends to an altitude of ~300 km before exiting the atmosphere (also a function of atmosphere density) [6]. Two hours after aerocapture a propulsive maneuver raises pericenter out of the atmosphere and circularizes the orbit. For an entry velocity of 6.5 m/s the pericenter-raise Δv is 150 m/s (deterministic). The target and orbit characteristics (after circularization) are provided in Table 1.

Table 1: Orbiter Entry Interface Target and Post-insertion Orbit Characteristics
Entry time: Aug 20, 2016 23:56:18 UTC

EI Target				Orbit Characteristics		
Altitude (km)	Latitude (deg)	eFPA (deg)	Entry Velocity (km/s)	Altitude (km)	Inclination (deg)	Period (hours)
1000	80N	-36.8	6.5	1700	100	5

NAVIGATION DATA

Optical Data
Optical data consists of digital images of Titan, other Saturnian satellites, and possibly Saturn, set in front of a stellar background. Background stars, combined with planetary ephemerides, establish the spacecraft-Titan relative position by astrometry. And since radiometric data are unable to resolve a gravity signature from Titan until a spacecraft is nearly upon it, target imaging is important for approach navigation.

The optical navigation campaign begins at E-75 days. (In practice, it's more likely that navigation images would start as early as T-180 days.) Ground-based facilities will process transmitted pictures to extract the optical observables; then the data will be combined with radiometric measurements.

Opnav transmissions will be constrained by the downlink data rate (600 bps). Probably more significant, however, are conflicts with competing spacecraft activities. Anticipating such demands during the last fortnight of cruise, a maximum rate of one opnav per every 2-4 hours was selected as a reasonable compromise [7].

At the beginning of the approach phase, one picture every other day is shuttered. (Titan's orbital period is 16 days.) The picture frequency increases to five per day, and increases again to approximately ten opnavs per day within 16 days of Titan. This yields approximately 200 images in the optical data set.

The imaging system envisioned here follows a design similar to the Mars Reconnaissance Orbiter optical navigation camera. Relevant technical specifications of the advanced (MRO) camera are: aperture = 6 cm, focal length = 50 cm, field-of-view = 1.4° per side, detector = 1024x1024 CCD array, pixel resolution = 50 μrad, mass = 2.7 kg, peak power = 4 W [8].

For comparison, the Cassini wide-angle camera has these specifications: aperture = 6 cm, focal length = 20 cm, field-of-view = 3.5° per side, detector = 1024x1024 CCD array, pixel resolution = 60 μrad, mass = 27 kg, peak power = 35 W [7]. The advanced (MRO) camera offers higher resolution yet weighs less and is more energy efficient.

See Appendix 1 for optical data weights.

Tracking Data

The baseline navigation data set throughout the mission consists of two-way coherent X-band doppler and two-way coherent X-band range measurements. These data are augmented during approach with optical observations and ΔVLBI measurements.

ΔVLBI data enhances the navigation solution relative to that achievable with doppler, range and optical data (although it is the optical data that dominates in a ranking of the relative importance of the four data types). In general ΔVLBI data, or specifically for this analysis Delta Differenced One-way Range (ΔDOR), has limited effectiveness because of the spacecraft's distant location in the solar system, but it can be used to great advantage in combination with other data types.

Table 2: Doppler and Range Tracking Coverage

Start	End	Coverage
E-75days	E-41	3 tracks/week
E-40	E-11	2 tracks/day
E-10	Entry	3 tracks/day

Data frequencies used in this analysis for doppler, range and ΔDOR are provided in Tables 2 and 3. Data weights are provided in Appendix 1.

Table 3: DDOR Coverage

Start	End	Observation Frequency
E-75days	E-51	1 per week
E-50	E-35	3.5 per week
E-34	Entry	14 per week

Navigation Performance

Significant error sources in the navigation model are listed below. (Also see Appendix 1 which lists all error sources and *a priori* uncertainties.) The combined effect of maneuver execution errors and orbit determination errors mapped to the atmospheric entry interface point is called the delivery accuracy.

Ephemeris Determination

Bounds on the Titan ephemeris errors in the year 2016 are given in Table 4. The tabulation is given in a Saturn-centered RTN coordinate system, where R represents the radial direction, T the down-track direction and N the out-of-plane direction.

Significant improvements to the Titan ephemeris between now and 2016 should occur. The suffixes BOC and EOC in Table 4 illustrate this improvement with respect to current ground-based observational knowledge (indicated by "GB"). "BOC" (beginning of Cassini) denotes Titan's ephemeris knowledge shortly after Cassini's arrival at Saturn. "EOC" (end of Cassini) represents the expected Titan ephemeris knowledge in 2008 after approximately 44 Cassini flyby encounters. The improvement is appreciable.

Table 4: Titan, Saturn and Earth Ephemeris Uncertainties (3σ)

Mapped to August 21, 2016

Central Body	R (km)	DT (km)	OOP	RSS
Earth	0.01	2.6	3.9	4.7
Saturn GB*	95	405	135	437
Saturn BOC	27	81	51	99
Titan GB*	120	570	300	655
Titan BOC	6	482	6	492**
Titan EOC	1	30	1	104**

* Representative of accuracies currently possible from ground-based observations. Not mapped to 2016.
** Combined with Saturn BOC error.

The BOC and EOC ephemeris accuracies in Table 4 include secular degradation to 2016.

Maneuver Determination

Maneuvers scheduled during the approach phase are listed below. Table 5 defines the baseline targeting strategy adopted for this analysis (last maneuver at E-24 hours), while Table 6 illustrates an alternative targeting strategy (last maneuver at E-6 hours). The important difference between the two strategies is the probe release time. For the strategy outlined in Table 5, the probe is released at E-30 days whereas the alternative strategy releases the probe at E-5 days. (A late release time, *i.e.* E-5 days, is preferred in order to reduce probe instability growth after separation and achieve an accurate delivery.)

The TCM3 magnitude, parameterized as a function of probe release time, is plotted in Figure 3. The remaining maneuvers in Tables 5 and 6 are small statistical maneuvers. The magnitudes of these clean-up maneuvers have not been analyzed, but the mean Δv for TCM1 and TCM2 probably will not exceed 1 m/s each for either strategy. For the PR-30 strategy, the combined total of TCM4 and TCM5 will be on the order of 5-10 m/s (at a confidence level of 1σ). For the PR-5 strategy, the mean Δv for TCM4 and TCM5 will be much greater.

Table 5: PR-30 Approach Phase TCMs

TCM*	Time**	OD Data Cutoff**	Description
TCM-1	E -60 days	E - 65 days	Correct SEP cruise errors; penultimate probe targeting.
TCM-2	E -31 days	E - 35 days	Final probe targeting to probe-entry aim point.
PR	E -30 days	E – 30 days	Probe release.
TCM-3 (TDM)	E -29 days	~E – 30 days	Deflect orbiter to entry interface point.
TCM-4	E – 9 days	E – 11 days	Correct deflection maneuver errors; penultimate targeting.
TCM-5	E - 24 hrs	E – 48 hrs	Final orbiter targeting to orbiter-entry aim point

*Numbered starting at the beginning of the approach phase.
**With respect to orbiter entry (E) time.

The 1σ execution errors assumed here are: fixed magnitude error of 1 mm/s, propor-tional magnitude error of 1% per axis, fixed pointing error of 1 mm/s per axis, and proportional pointing error of 2 milliradians per axis for the deflection maneuver and 1 milliradians per axis for the other TCMs.

Orbit Determination

The dominant orbit determination uncertainties consist of the reaction caused by the probe release mechanism, ephemeris errors, TCM execution uncertainties, and data errors.

Table 6: PR-5 Approach Phase TCMs

TCM	Time	OD Data Cutoff	Description
TCM-1	E –60 days	E - 65 days	Correct SEP cruise errors; penultimate probe targeting.
TCM-2	E -6 days	E - 8 days	Final probe targeting to probe-entry aim point.
PR	E -5 days	E – 5 days	Probe release.
TCM-3 (TDM)	E -4 days	~E – 5 days	Deflect orbiter to entry interface point.
TCM-4	E -24 hrs	E – 48 hrs	Correct deflection maneuver errors; orbiter targeting.
TCM-5	E -6 hrs	E – 15 hrs	Final orbiter targeting to orbiter-entry point.

Results

Delivery errors are a combination of orbit determination errors and maneuver execution errors. Results for both the orbiter bus and entry probe are presented here. Unless noted otherwise, all results assume EOC Titan ephemeris knowledge. Sensi-tivities examined were: delivery improve-ments due to optical navigation data and/or ΔDOR observations, and entry FPA uncertainties parameterized by probe release time and/or the orbiter over-flight delay.

Orbiter Vehicle. Mission OD uncertainties in the Titan B-plane for a probe release time of E-30 days and an over-flight delay of 3.5 hours are shown in Figure 5. TDM is not shown in Figure 5; it lies off the left edge of the figure. But it's the rapid reconstruction of TDM (*i.e.* falling FPA uncertainty shown in Figure 5) that enables aerocapture to be undertaken.

Sensitivities are shown in Tables 7 - 10. Table 7 illustrates the improvement in the overall delivery accuracy due to the incremental addition of advanced data types to the basic doppler and range data set. The results in Table 8 show the delivery sensitivity to probe release time. Table 9 shows the effect of varying the *a priori* Titan ephemeris. Table 10 lists the probe delivery accuracy.

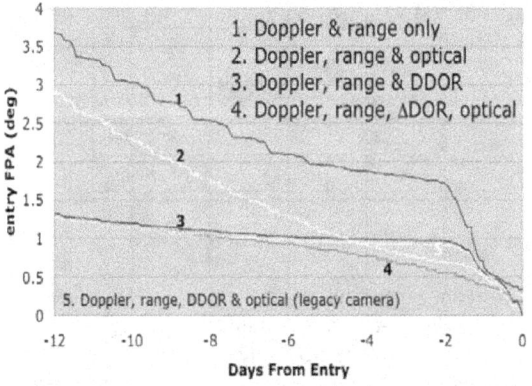

Figure 5: FPA v. Time-to-Go (3σ)

Margin remains in the navigation sub-system. Table 7 (and Figure 5) show the delivery requirement is satisfied by any tracking option, excluding the doppler and range only option.

The probe release time influences the magnitude of TDM and the telecom link between the probe and bus. Greater orbiter eFPA uncertainties can be expected for large distances between the orbiter and probe (*i.e.* long over-flight delay times) and for late probe release times (*i.e.* releasing closer to Titan).

This performance is illustrated in Table 8.

Knowledge of Titan's orbital position increases during Cassini's tour (Table 4). This improvement is apparent in the flight path angle errors shown in Table 9.

**Table 7: Orbiter Delivery Accuracy (3σ)
Data Sensitivity**

Probe release = E-30 days, Over-flight delay = 3.5 hours				
	Doppler & Range Only	Doppler Range & ΔDOR	Doppler Range & Optical	Doppler Range Optical ΔDOR
Preliminary Reqm't	±1.0	±1.0	±1.0	±1.0
Nominal Delivery				
Data Cutoff at E-48 hours				
Semi-major axis (km)	68.7	39.1	28.1	23.3
Semi-minor axis (km)	17.4	16.0	13.3	11.7
Ellipse angle (deg)	75.2	66.0	64.9	53.7
Entry time (s)	17.5	9.1	6.7	4.8
B magnitude (km)	68.7	38.6	27.6	22.0
Flight Path Angle (deg)	±1.7	±1.0	±0.7	±0.6
Alternate Delivery				
Data Cutoff at E-30 hours				
Semi-major axis (km)	37.0	30.0	21.9	20.9
Semi-minor axis (km)	16.1	14.2	12.6	10.9
Ellipse angle (deg)	65.3	55.0	57.4	48.0
Entry time (s)	10.9	7.9	5.5	4.0
B magnitude (km)	36.5	28.5	21.1	19.2
Flight Path Angle (deg)	±0.9	±0.7	±0.5	±0.5
Parameter Update				
Data Cutoff at E-3 hours				
Semi-major axis (km)	15.8	7.2	13.1	7.1
Semi-minor axis (km)	1.5	1.6	1.5	1.7
Ellipse angle (deg)	56.2	57.2	56.3	57.9
Entry time (s)	3.4	1.3	1.5	1.2
B magnitude (km)	14.9	6.9	12.3	6.7
Flight Path Angle (deg)	±0.4	±0.2	±0.3	±0.2

**Table 8: Orbiter Delivery
Probe Release Time v. Over-flight Delay
Entry FPA Uncertainty (3σ)**

	Probe Release E – 30 days		Probe Release E – 5 days	
Over-flight Delay ->	3.5 hrs	2 hrs	2 hrs	1 hr
E – 48 hours (deg)	±0.6	±0.6	±1.1	±1.0
E – 30 hours (deg)	±0.5	±0.4	±0.9	±0.8
E – 3 hours (deg)	±0.2	±0.2	±0.6	±0.6

**Table 9: Orbiter Delivery
Probe Release Time v. Ephemeris Sensitivity
Entry FPA Uncertainty (3σ)**

Probe Release = E - 30 days Over-flight Delay = 3.5 hours			
Titan ephemeris ->	circa 2008 (EOC)	circa 2004 (BOC)	circa 2000 (GB)
E – 48 hours (deg)	±0.6	±0.7	±3.6
E – 30 hours (deg)	±0.5	±0.6	±3.5
E – 3 hours (deg)	±0.2	±0.3	±0.5

The delivery dispersions of entry interface states resulting from monte carlo simulations of navigation errors are shown in Figure 6. The larger scatter (grey points) represents the B-plane uncertainty at the E-2 day delivery, while the tighter scatter (black points) depicts the uncertainty at an E-3 hour delivery.

The E-2 day scatter shown in Figure 6 maps into flight path angle uncertainty shown in Figure 7. Figure 7 plots FPA as a function of entry velocity for the E-2 day delivery. The scatter is not clustered around -36.8° because the navigation entry states are mapped to 1 minute before EI (radius = 3821 km) and not the EI (radius = 3575 km).

Earlier Study. A navigation analysis undertaken in May 2002 generated preliminary results for the Titan aerocapture study group, as discussed in Reference 1. The results presented here represent more recent analyses. (In the earlier study, although the advanced camera was also investigated, it did not represent the baseline case. Ephemeris revisions have also occurred since then.)

Probe capsule. The ultimate probe-targeting maneuver occurs thirty-one days before entry (in the baseline case). Subsequent to that maneuver the spacecraft spins up to prepare for probe release. At E-30 days the probe separates from the spacecraft bus. Contact with the probe is broken for 30 days while the probe follows a ballistic path to Titan. Upon reaching the entry interface point (radius equal to 3575 km) the probe begins broadcasting telemetry on UHF. The orbiter, at that time about 80,000 km behind the probe, begins relaying the probe's data to Earth for the next 3.5 hours. Contact with the probe is lost at about the time the orbiter reaches its entry interface point.

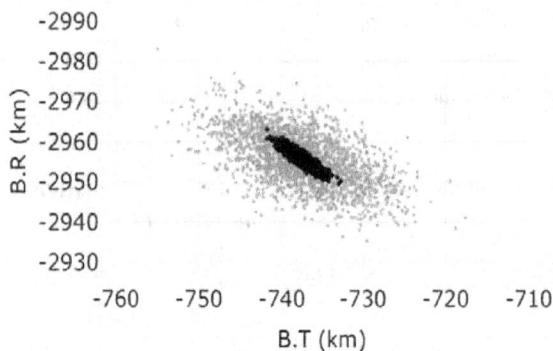

Figure 6: E-2 day and E-3 hour Deliveries Mapped to Entry Interface Point*

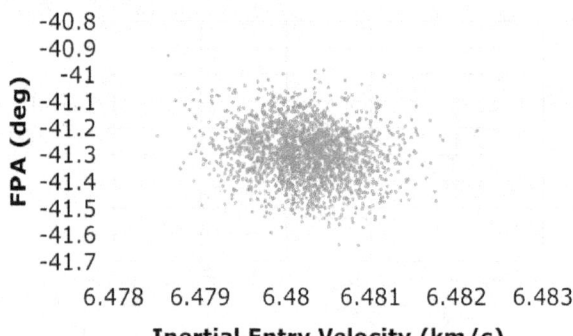

Figure 7: FPA v. Entry Velocity at EI*

The requirement to deliver the probe to Titan can be satisfied, as shown in Table 10, for either an E-30 day release or an E-5 day release. The latter probe release time delivers the probe to Titan with higher accuracy. The probe can separate as early as ~E-90 days and still meet the delivery requirement.

Table 10: Probe Delivery Accuracy (3σ) Probe Release Time and Camera Sensitivity

Camera Design	Probe Release E – 30 days		Probe Release E – 5 days	
	circa 2008	legacy camera	circa 2008	legacy camera
Preliminary Requirement	±5.0	±5.0	±5.0	±5.0
Flight Path Angle (deg)	±1.9	±1.9	±0.8	±0.9

The probe model differs in minor ways from the bus model. The significant differences to note are the initial state uncertainties due to the probe-release mechanism, the absence of any trajectory correction maneuvers after separation, and the paucity of tracking data after separation.

DISCUSSION

Significant benefits accrue by assuming Titan's position will be known to about the same level of certainty in 2021 as will exist at the time of Cassini's last flyby of Titan. This is a reasonable assumption since Titan's orbit has few perturbations and over a decade any error growth should be small (unlike the Galilean satellites of Jupiter). The sensitivity of the delivery to Titan's ephemeris is shown in Table 9.

Optical data is an important component in the navigation data suite, although its relative utility diminishes given precise and accurate Titan position knowledge. The EOC Titan uncertainty in Table 4 is nearly equal to the accuracy achievable with optical navigation (within a factor of ~2). Therefore opnavs contribute proportionally less to navigation early in the approach phase because errors other than Titan's ephemeris dominate (*e.g.* error in the spacecraft state). See plot 2 in Figure 5. In addition to a measurable improvement to the delivery from incorporating the Cassini EOC-level Titan ephemeris, modest improvements in camera technology can also improve the delivery (the difference between plots 4 and 5 in Figure 5).

Optical navigation images of Titan, because of its atmosphere, have relatively large uncertainties, especially in the week before entry. This uncertainty is partially mitigated by incorporating navigation pictures of the small, icy moons of Saturn.

ΔDOR improves delivery accuracy (see Figure 5 and Table 7). This improvement in delivery accuracy comes about because the *a priori* Titan position uncertainty is less than the optical data noise (early in the approach phase). This means delivery knowledge is not dominated by ephemeris errors, but rather by the spacecraft's state uncertainty. ΔDOR observations decrease spacecraft errors. Only within four days of Titan does optical data dominate the solution.

ΔDORs and optical data are orthogonally complementary and thus can combine to yield plot 4 in Figure 5. This is the best delivery in the current study. Note that plot 5 represents a similar simulation, but with a less advanced camera. In the latter case optical data noise swamps ephemeris and spacecraft errors, resulting in less precise orbit determination.

There is no advantage to using Ka-band tracking in place of X-band. There may be benefit to Ka-band ΔDOR observations, but this parameterization has not yet been explored.

Separation of the probe from the orbiter can occur as early as approximately 90 days before Titan and satisfy the probe's entry requirements. However, at the opposite end, for a separation equal to E-5 days, several disadvantages become apparent. The maneuver schedule in the last week is compressed significantly with respect to a separation at E-30 days or earlier. The magnitude of the deflection maneuver (TDM) is large. This is offset with an accurate probe delivery to Titan.

The data arc used for the E-30 day probe release consists of 30 optical navigation images and 32 ΔDOR observations. For the E-30 day release time, both the baseline case (MRO-like camera) and a less capable legacy imaging system provide equivalent probe deliveries, both well within the delivery requirement.

Entry time uncertainty for the probe is 20 seconds (3σ) for the baseline case (*i.e.* for an E-30 day release, four navigation data types, EOC ephemeris). The relay link on the orbiter bus needs this margin for the telecom design.

The Titan orbiter mission is boosted by SEP during early cruise, but during the latter half of the mission conventional thrusters control the spacecraft. As a rough estimate of propellant loading, for an E-30 day separation and a 3.5 hour over-flight delay, the spacecraft will need to carry propellant sufficient to perform about 300 m/s of velocity change, including both deterministic and statistical maneuvers (up to and including the post-insertion orbit circularization burn, but not including on-orbit maintenance propellant) [2].

Entry FPA results or expected results from other missions are summarized in the table below. (MER, Huygens, and Stardust have not yet arrived at Mars, Titan and Earth respectively at the time of this writing.)

MPL and Huygens stand out in the short list above with high uncertainties.

The MPL mission was characterized by unbalanced and mis-modeled thrusting activities. The level of thrusting required by the ACS system to maintain attitude significantly exceeded pre-launch expectations, and this mis-modeling contributed to the entry flight path angle uncertainty shown in Table 11.

Table 11: Delivery Accuracy Comparison (3σ) Orbiter and Probe v. Other Missions

	Entry FPA	Error	Delivery Time	Reqm't
Titan orbiter	-36.8°	±0.6°	E-2 d	<±1.0>
Mars Pathfinder	-14.2°	±0.4°	E-2 d	±1.0
MPL	-12.0°	±1.0°	E-2 d	~±0.5
MER	-11.5°	±0.2°	E-2 d	±0.25
Stardust	-8.2°	~±0.8°	E-2 d	±0.8
Titan probe	-50.0°	±1.9°	E-35 d	±5.0
Galileo probe	-8.6°	±0.6°	E-140 d	±1.4
Huygens probe	-64.0°	±3.0°	E-21 d	±3.4

<-> indicates a proposed requirement

Huygens will be released from Cassini 21 days before entry and, with an uncertainty of +/-3.0°, is not significantly different from the Titan orbiter probe delivery. The Huygens delivery is larger than the Titan probe because it does not have access to the Titan EOC ephemeris. Another reason is the tour re-design Cassini has undergone recently [9]. The release time is no longer as favorable for probe delivery, subjecting Huygens to additional perturbations.

The MER delivery, on the other hand, is significantly smaller than MPL and the current delivery. The reasons include: well known Mars' ephemerides, large gravity-well, lack of a TDM, and no force mis-modeling.

SUMMARY

This study has baselined the use of optical observations and ΔDOR measurements in addition to doppler and range data for delivering a probe and orbiter to Titan, and has assumed a Cassini-era Titan ephemeris. The orbiter can satisfy its proposed entry FPA error requirement of ±1.0° (3σ) as early as 7 days from Titan. By E-2 days that error has decreased by 170% (*i.e.* to ±0.6°). Nav-igation is sensitive to ΔDOR measurements, and the E-2 day delivery accuracy improves by 0.1° vis-à-vis only doppler, range and optical data, which is significant. An alternate delivery at E-30 hours reduces the E-2 day FPA uncertainty by an additional 15%.

Current camera technology (Cassini-era) is sufficient to meet the delivery requirement. More advanced cameras (MRO-era) will significantly improve the delivery and offer mass savings.

The orbiter cannot meet preliminary entry requirements using a pre-Cassini ground-based *a priori* Titan ephemeris such as that described in Table 4.

The pericenter-raise maneuver following orbit insertion is a significant issue that has not been addressed here. Maneuver autonomy may be required.

This work represents a first-cut effort at determining concept feasibility. Many simplifying assumptions were made, especially with respect to the optical data, in order to accomplish this study in a timely manner.

ACKNOWLEDGEMENTS

This work was performed at the Jet Propulsion Laboratory, California Institute of Technology under contract from NASA. The work was funded through the In-space Propulsion program.

The author would like to thank J.L. Hall, M. Noca, R. Bailey, T. Spilker, W.M. Owen, T. Goodson, P. Antreasian, D. Roth, P. Knocke, and L.A. Cangahuala for helpful comments and support.

Many thanks also to R. Ionasescu for providing the simulated Cassini-era Titan ephemeris. Carl Sauer constructed the intial SEP trajectory between Earth and Saturn.

REFERENCES

1. Lockwood, M.K., "Titan Aerocapture Systems Analysis", paper AIAA-2003-4799, Joint Propulsion Conference, Huntsville Alabama, July 21-23, 2003.
2. Bailey, R.W., J.L. Hall, T. Spilker, "Titan Aerocapture Mission and Spacecraft Design Overview", paper AIAA-2003-4800, Joint Propulsion Conference, Huntsville Alabama, July 21-23, 2003.
3. Noca, M., JPL, personal communication, May 2003.
4. Lockwood, M.K., D.W. Way, Langley Research Center, personal communication, March 2002.
5. Lorenz, R.D. et al, "The Seas of Titan", EOS, Vol. 84, Number 14, 8 April 2003.
6. Justus, C.G., "Aerocapture Systems Analysis Review", viewgraph presentation, 29 Aug. 2002.
7. Owen, W.M., JPL, personal communications, April 2002 - May 2003.
8. Fraschetti, G., JPL, "MRO Optical Navigation Camera Preliminary Design Review", viewgraph presentation, 8 May 2002.
9. Strange, N., T. Goodson, Y. Hahn, "Cassini Tour Redesign for the Huygens Mission", paper AIAA-2002-4720, AIAA Specialist Conference, Monterey Calif, 5-8 August 2002.

Appendix 1: *A Priori* Navigation Model Uncertainties

Error Source	A Priori Uncertainty (1σ)	Correlation Time	Comments
Data			
doppler (mm/s)	0.1	-	X-band
range (m)	20	-	Relatively high uncertainty in lieu of range biases
ΔDOR (nrad)	4	-	0.11 ns (X-band)
optical (pixels)	0.25 - ~5.0	-	Larger error corresponds to smaller range
Estimated Parameters			
epoch state			
position (km)	1000	-	
velocity (km/s)	1	-	
Saturn ephemeris (km)	(9, 27, 17)	-	R,AT,OOP, circa 2016 (5% DE405 error)
Saturn mass (km^3/s^2)	0.0002	-	circa 2016 (from simulated ephemeris)
Saturn pole direction (mdeg)	(0.12, 0.14)	-	R.A., Dec, circa 2016 (from simulated ephemeris)
Titan ephemeris (km)	(1, 10, 1)	-	R, AT, OOP, circa 2016 (from simulated ephemeris)
camera pointing error (deg)	(0.5, 0.5, 2)	0	R.A., Dec, Twist; estimated per observation
non-gravitational accelerator (km/s^2)	1.0×10^{12}	10 days	Spherical covariance, estimated daily (1 day batches)
solar pressure (%)	10	-	Reflectivity coefficient
ACSΔV (mm/s), 1 every 3 wks.	(2, 2, 2)	-	(line-of-sight, lateral, normal) components
TCMs (mm/s)			spherical covariance
TCM-1	4	-	3% (3σ) proportional error (per axis)
TCM-2	2	-	3 mm/s (3σ) fixed error (per axis)
proble release	5	-	Probe release at E – 30 days
TCM-3	330	-	TCM3 at E – 29 days
TCM-4	5	-	
TCM-5	7	-	TCM4 at E – 9 days; TCM5 at E – 1 day
Earth pole direction (cm)	2→10	0	(X and Y). Ramps to higher value during final week of data.
UT1 (cm)	2→10	0	(For UT1, ~10cm -> 0.26 ms.)
ionosphere-day (cm)	55	0	S-band values
ionosphere-night (cm)	15	0	
troposphere (cm)	1	0	
Considered Parameters			
station locations (cm)	3	-	
quasar locations (nrad)	2	-	for ΔDOR data

Appendix 2: *B*-plane Description

Planet or satellite approach trajectories are typically described in aiming plane coordinates referred to as "*B*–plane" coordinates (see Figure). The B-plane is a plane passing through the body center and perpendicular to the asymptote of the incoming trajectory (assuming two body conic motion). The "B-vector" is a vector in that plane, from the body center to the piercing-point of the trajectory asymptote. The B-vector specifies where the point of closest approach would be if the target body had no mass and did not deflect the flight path. Coordinates are defined by three orthogonal unit vectors, **S**, **T**, and **R**, with the system origin at the center of the target body. **S** is parallel to the spacecraft v_∞ vector (approximately the velocity vector at the time of entry into the target body's gravitational sphere of influence). **T** is arbitrary, but typically specified to lie in the ecliptic plane (the mean plane of the Earth's orbit), or in the body equatorial plane. Finally, **R** completes an orthogonal triad with **S** and **T**.

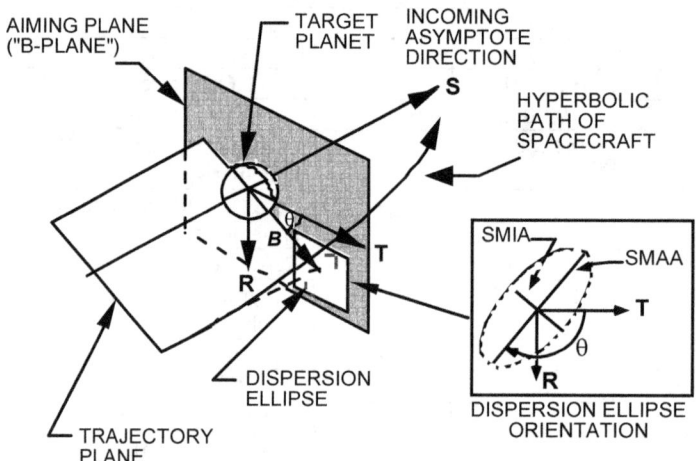

Fig. Aiming Plane Coordinate System Definition

Orbit determination errors can be characterized by a statistical dispersion ellipse in the aiming plane (*B*–plane) and a statistical uncertainty along the **S** (down-track) direction. In the Figure, SMIA and SMAA denote the semi–minor and semi–major axes of the dispersion ellipse (*i.e.* 50% of the distance across the ellipse along the respective coordinate). The angle θ is measured clockwise from **T** to SMAA.

ENGINEERING-LEVEL MODEL ATMOSPHERES FOR TITAN AND NEPTUNE

C.G. Justus and Aleta Duvall
Computer Sciences Corporation, Huntsville, AL

D.L. Johnson
NASA Marshall Space Flight Center, Marshall Space Flight Center, AL

ABSTRACT

Global reference atmospheric models for Titan and Neptune have been newly developed for utilization in NASA's aerocapture systems analysis studies. Their applicability to the engineering design of guidance, navigation and control, and thermal protection systems is discussed. The legacy and key features of Titan-GRAM and Neptune-GRAM are presented, with emphasis on the characterization of quasi-random atmospheric density perturbations. Sample Monte Carlo output for each model is presented.

INTRODUCTION

Engineering-level atmospheric models for Titan and Neptune have been newly developed for use in NASA's systems analysis studies of aerocapture applications in missions to the outer planets. Analogous to highly successful Global Reference Atmospheric Models for Earth (GRAM)[1] and Mars (Mars-GRAM)[2,3], the new models are called Titan-GRAM and Neptune-GRAM. Like GRAM and Mars-GRAM, an important feature of Titan-GRAM and Neptune-GRAM is their ability to simulate quasi-random perturbations for Monte Carlo analyses in developing guidance, navigation and control algorithms, and for thermal systems design. Figure 1 compares density-height profiles for Earth, Mars, Titan, and Neptune. Relatively low scale heights (~10 km) make densities for Earth and Mars drop rather rapidly with altitude. Significantly higher scale height values for Titan and Neptune (~40 km) make these atmospheres considerably "thicker". Titan's large density scale height is due to its low gravity (~0.14 x

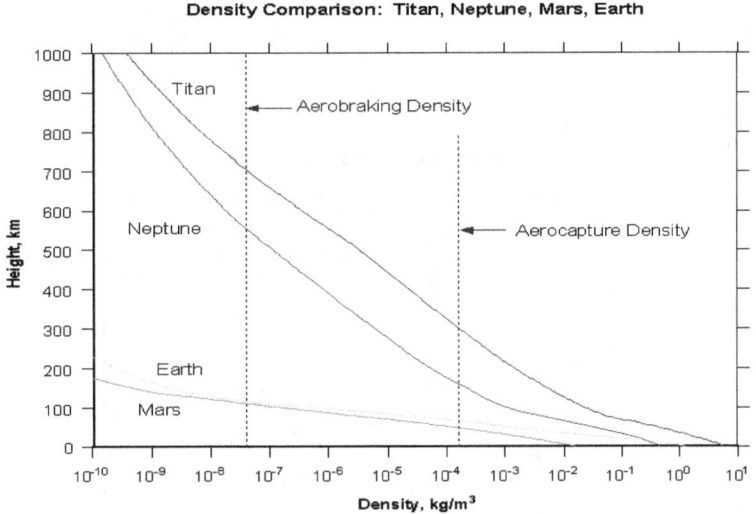

Figure 1 - Typical density versus altitude on Earth, Mars, Titan, and Neptune.

Earth gravity), while that for Neptune is due to its low atmospheric mean molecular weight (~2.3 versus ~29 for Earth). Vertical dotted lines in Figure 1 show density values and altitudes at which aerocapture or aerobraking maneuvers would occur on these planets.

BASIS FOR THE ATMOSPHERIC MODELS

In GRAM and Mars-GRAM, input values for date, time, latitude, longitude, etc. are used to calculate planetary position and solar position, so that effects of latitude variation, and seasonal and time-of-day variations can be computed explicitly.

A simplified approach is adopted in Titan-GRAM and Neptune-GRAM whereby these effects (as well as effects of relatively large measurement uncertainties for these planets) are represented within a prescribed envelope of minimum-average-maximum density versus altitude. Figure 2(a) shows this envelope for Titan, for which engineering atmospheric profiles of Yelle et al.[4] are used. For Neptune, data from Cruikshank[5] were employed to generate a comparable envelope, shown in Figure 2(b).

A single model input parameter (Fminmax) allows the user of Titan-GRAM or Neptune-GRAM to select

(a)
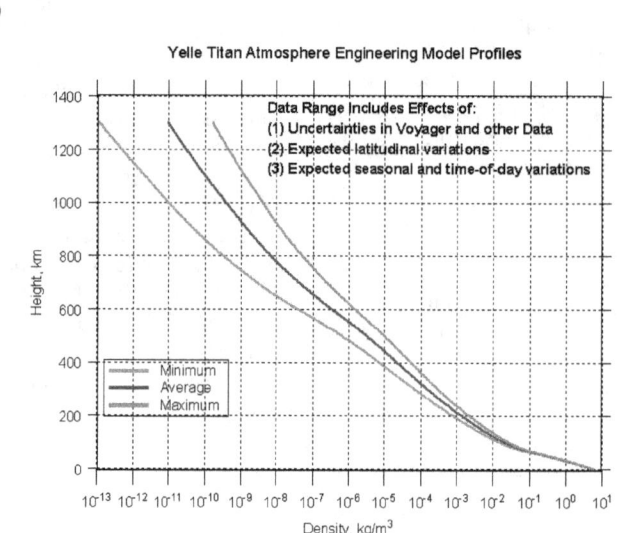

(b)
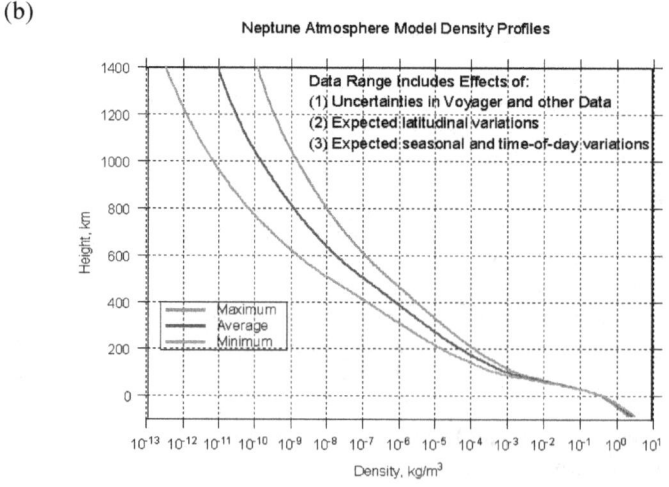

Figure 2 – Minimum, average, and maximum density versus altitude for (a) Titan[4] and (b) Neptune (developed from data in Cruikshank[5]).

where within the min-max envelope a particular simulation will fall. Fminmax = -1, 0, or 1 selects minimum, average, or maximum conditions, respectively, with intermediate values determined by interpolation (i.e. Fminmax between 0 and 1 produces values between average and maximum). Effects such as variation with latitude along a given trajectory path can be computed by user-selected representations of variation of Fminmax with latitude.

Yelle et al. assumed methane mole fractions of 5%, 3%, and 1% for minimum, average, and maximum density Titan atmospheres, respectively. Aerothermal analyses (other papers in this session) showed very strong radiational heating by CN radicals produced from methane and nitrogen reactions in the high-temperature aerocapture entry shock layer. Because of the importance of methane concentration in this process, an option was added in Titan-GRAM to allow the user to select any amount of methane (up to 5%), while retaining the original Yelle et al. profiles of mean molecular weight versus height and density versus height. Atmospheric density perturbations are computed by methods similar to those used in GRAM and Mars-GRAM. Perturbation magnitudes in Titan-GRAM and Neptune-GRAM are estimated from a methodology similar to that of Strobel and Sicardy[6], based on expected wave saturation effects.

In particular, perturbation magnitudes are modeled from an approximate fit to the wave saturation condition relation

$$\rho'_{max}/\rho_0 = [L/(2\pi H)](1 + [L/(2\pi H)]^2)^{1/2} (dT_0/dz + g/C_p)/(g/R)$$

where H is pressure scale height ($R T_0 / g$), R is gas constant, g is acceleration of gravity, T_0 is background mean temperature, and C_p is specific heat at constant pressure. Model perturbation magnitudes are illustrated in Figures 3(a) and 3(b).

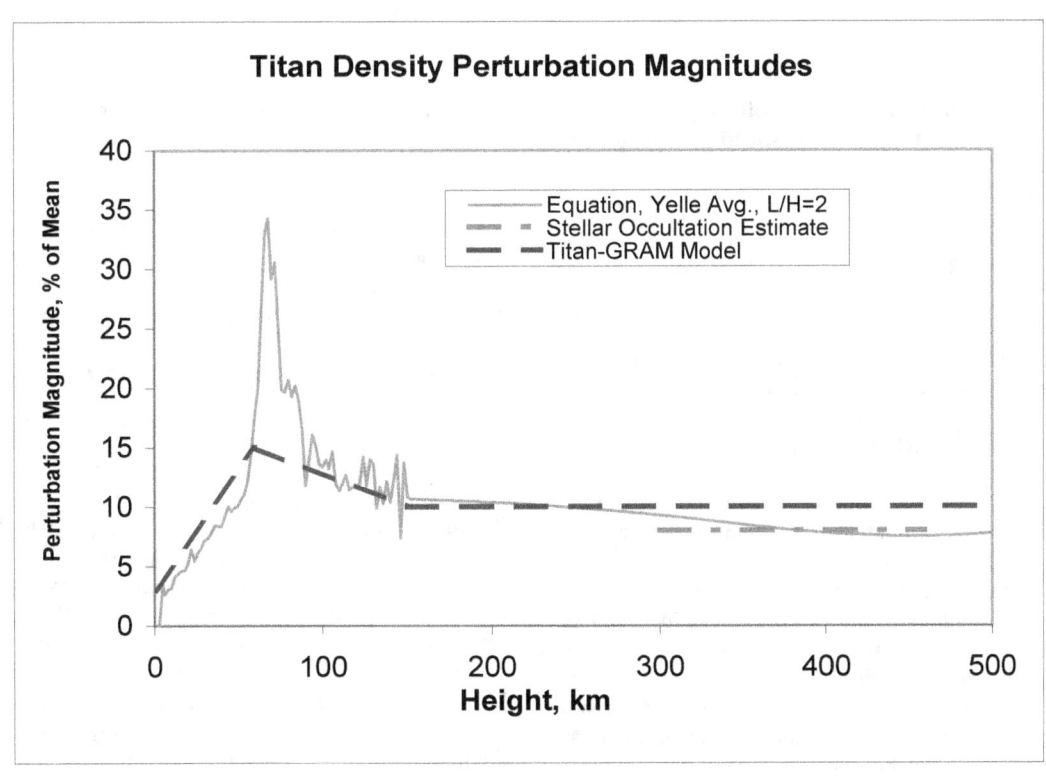

Figure 3(a) – Perturbation magnitudes in Titan-GRAM. Model values approximate equation (1), with ratio L/H=2 and dT_0/dz from respective average atmospheres. Textured line is estimated value from Figure 1(a) of Strobel and Sicardy[6].

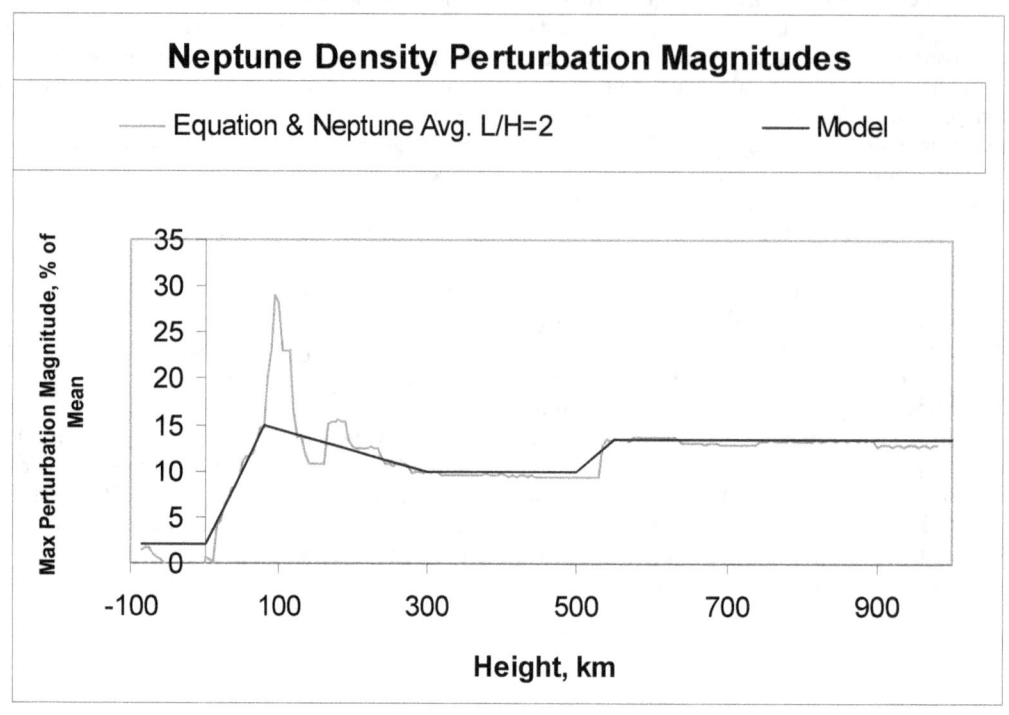

Figure 3(b) – Perturbation magnitudes in Neptune-GRAM. Model values approximate equation (1), with ratio L/H=2 and dT_0/dz from respective average atmospheres.

SAMPLE MODEL RESULTS

Sample Monte Carlo outputs from Titan-GRAM and Neptune-GRAM are shown in Figure 4.

CONCLUSIONS

Titan-GRAM and Neptune-GRAM are engineering-level atmospheric model for Titan and Neptune, suitable for a wide range of mission design, systems analysis, and operations tasks. For orbiter missions, Titan-GRAM and Neptune-GRAM applications include analysis for aerocapture or aerobraking operations, analysis of station-keeping issues for science orbits, analysis of orbital lifetimes for end-of-mission planetary protection orbits, and atmospheric entry issues for accidental break-up and burn-up scenarios. For Titan lander missions, Titan-GRAM applications include analysis for entry, descent and landing (EDL), and guidance, navigation and control analysis for precision landing. Using Titan-GRAM or Neptune-GRAM perturbation models in Monte Carlo mode make them especially suited for design and testing of guidance, navigation, and control algorithms and for heat loads analysis of thermal protection systems.

ACKNOWLEDGEMENTS

The authors gratefully acknowledge support from the NASA/Marshall Space Flight Center In-Space Propulsion Program. Particular thanks go to Bonnie James (MSFC), Manager of the Aerocapture Technology Development Project, to Michelle M. Munk (LaRC/MSFC), Lead Systems Engineer for Aerocapture, and to Ann Trausch and Melody Herrmann (MSFC), team leads and Mary Kae Lockwood (LaRC), technical lead for the Titan/Neptune Systems Analysis study. Model user feedback and suggestions from the following individuals are also greatly appreciated: Dick Powell, Brett Starr, and David Way (NASA LaRC), and Claude Graves and Jim Masciarelli (NASA JSC).

(a)

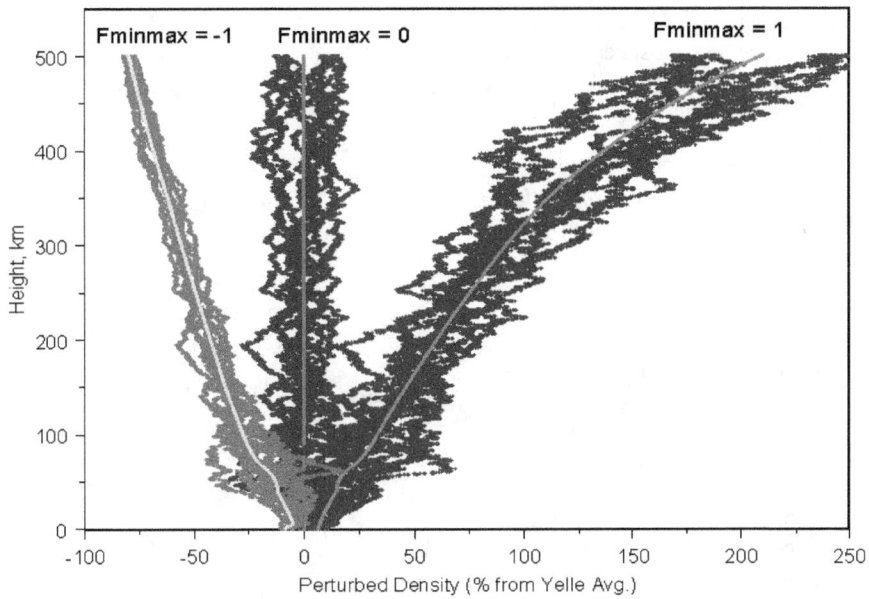

(b)

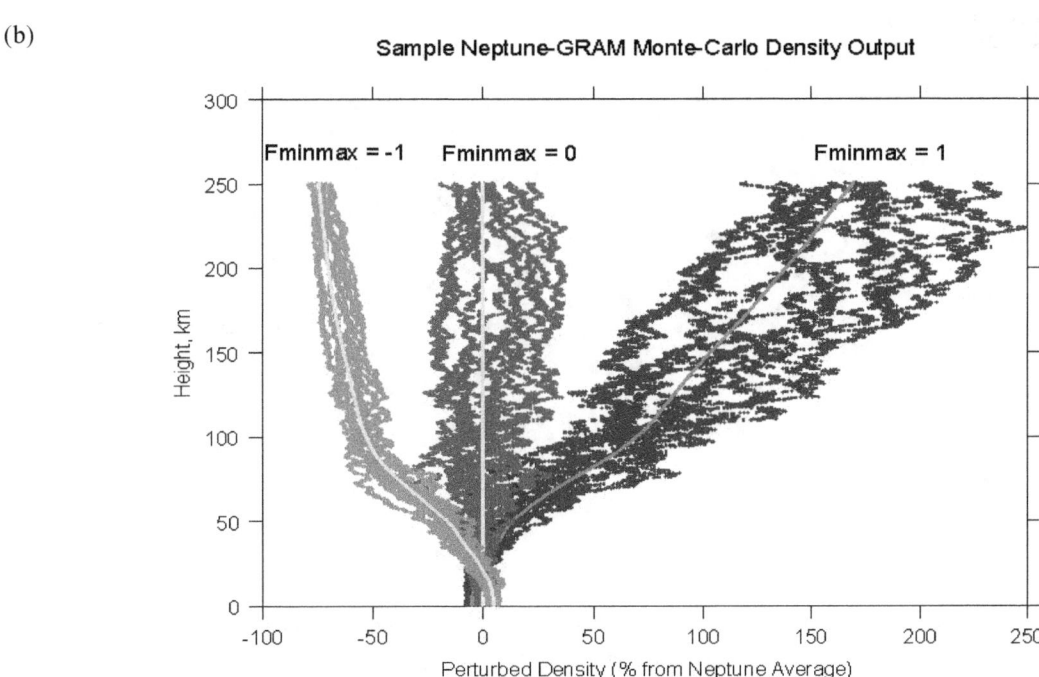

Figure 4 - **Sample Monte Carlo perturbation profiles from (a) Titan-GRAM and (b) Neptune-GRAM for Fminmax = -1, 0 and 1. Density values are expressed as percentage deviation from average density (Fminmax = 0) profile value.**

REFERENCES

[1] Justus, C.G., Johnson, D.L., and James, B.F., "New Global Reference Atmospheric Model (GRAM-99) and Future Plans", COSPAR 33rd General Assembly, Warsaw, Poland, Invited Paper C4.2-0003, July 2000.

[2] Justus, C.G., and Johnson, D.L., *Mars Global Reference Atmospheric Model 2001 Version (Mars-GRAM 2001) Users Guide*, NASA/TM-2001-210961, April 2001.

[3] Justus, C.G., James, B.F., Bougher, S.W., Bridger, A.F.C., Haberle, R.M., Murphy, J.R., and Engel, S., "Mars-GRAM 2000: A Mars Atmospheric Model for Engineering Applications", *Advances in Space Science*, Vol. 29, No. 2, 2002, pp. 193202.

[4] Yelle, R.V., Strobel, D.F., Lellouch, E., and Gautier, D., "Engineering Models for Titan's Atmosphere", *Huygens Science, Payload and Mission*, ESA SP-1177, Aug 1997.

[5] Cruikshank, D.P., Ed., *Neptune and Triton*, University of Arizona Press, Tucson, 1995.

[6] Strobel, D.F., and Sicardy, B., "Gravity Waves and Wind Shear Models", *Huygens Science, Payload and Mission*, ESA SP-1177, Aug. 1997.

GUIDANCE ALGORITHMS FOR AEROCAPTURE AT TITAN

James P. Masciarelli
NASA Johnson Space Center
Houston, TX

Eric M. Queen
NASA Langley Research Center
Hampton, VA

ABSTRACT

An evaluation of two different guidance algorithms for aerocapture at Titan is presented for a reference mission. The evaluation includes guidance algorithm comparison to the theoretical optimum performance, as well as guidance response to variation in entry flight path angle, atmosphere density, and aerodynamic parameters. Monte Carlo simulation results are also presented. The results show that both algorithms, originally developed for aerocapture at Earth and Mars, perform well for aerocapture at Titan.

INTRODUCTION

NASA is investigating the use of aerocapture for orbit insertion at Titan, one of Saturn's moons.[1] An autonomous guidance algorithm will be required to perform this maneuver and achieve the desired orbit. The Hybrid Predictor-corrector Aerocapture Scheme (HYPAS), and the Terminal Point Controller (TPC) are two algorithms that have been developed for aerocapture guidance. Both algorithms have previously been tested extensively in simulations for aerocapture at Earth and Mars, but no investigation has been done on their performance at Titan. This paper explores the performance of these two algorithms for aerocapture at Titan to determine the feasibility of aerocapture for a reference mission.

The proposed Titan Explorer was used as the reference mission to assess guidance performance for aerocapture at Titan.[2] In the reference mission, the vehicle enters the Titan atmosphere at an altitude of 1000 km and an inertial velocity between 6.5 and 10.0 km/s, depending on where in its orbit Titan is encountered. The vehicle utilizes a blunt body aeroshell, with lift-to-drag ratio (L/D) of 0.25, and ballistic coefficient (m/C_DA) of 90 kg/m^2. Lift modulation by bank angle control is used to manage the aerocapture trajectory. The target orbit is circular, with an altitude of 1700 km, and inclination of 101.6 deg.

The paper begins with an overview of each guidance algorithm, and points to other references that describe the algorithms in more detail. The process used to analyze the guidance algorithms for the reference Titan aerocapture mission is then explained. Finally, the performance of the two algorithms is presented. A companion paper presents more detailed aerocapture performance results for the reference mission.[3]

OVERVIEW OF GUIDANCE ALGORITHMS

HYPAS

The HYPAS algorithm guides a lifting vehicle through the atmosphere to a desired exit orbit apoapsis and inclination or plane using bank angle control. The guidance uses an analytically derived control algorithm based on deceleration due to drag and altitude rate error feedback. Inputs to the guidance algorithm are the current vehicle position, velocity, sensed acceleration, and vehicle attitude. The algorithm outputs a commanded bank angle and the direction to bank from the current attitude. The guidance algorithm is adaptable to a wide range of initial state vectors, vehicle lift-to-drag ratios and ballistic coefficients, planetary atmospheres, and target orbits by changing a set of initialization constants. Furthermore, by tuning these constants, other trajectory constraints can be controlled such as maximum dynamic pressure, deceleration, heat rate, and the amount of the theoretical corridor captured.

A significant feature of the HYPAS algorithm is that no reference trajectories are computed prior to flight; all reference values are computed and updated during flight. This analytic, "on-the-fly" approach leads to efficient code, minimal data storage requirements, and minimal preflight effort. The current implementation of the guidance algorithm uses 313 source lines of Fortran code, including code required to process navigation data. There are 37 initialization constants, most of which are functions of the vehicle's lift-to-drag ratio and ballistic coefficient, the planet's atmosphere and gravity, and the target orbit conditions. The non-numerical, non-iterative scheme ensures fast and consistent execution times.

The HYPAS guidance algorithm consists of two phases. In the first phase, or capture phase, bank angle commands are generated to stabilize the trajectory and drive the vehicle toward equilibrium glide conditions, where lift, gravity, and centripetal forces are balanced. When the vehicle has decelerated to a specified velocity, the second phase, or exit phase, begins. In the exit phase, the velocity vector at atmospheric exit altitude is analytically predicted each guidance computation cycle. The bank angle command is then adjusted so that the velocity achieved at exit altitude will produce an orbit with the target apoapsis. This two-phase approach allows separate tuning of initialization constants to maximize robustness during capture and maximize performance during exit.

An inclination (or wedge angle if targeting a specific plane) dead-band that is a function of inertial velocity is used to target the desired orbit inclination (or plane). Whenever the inclination error exceeds this dead-band, a bank reversal is commanded. The direction to bank is selected through a series of tests that examine current velocity, angular distance to roll, difference between desired and measured altitude rate, and difference between desired and measured drag.

The HYPAS guidance algorithm has undergone extensive laboratory development and testing. The original version of the algorithm was developed for the Aeroassist Flight Experiment (AFE) program[4], and its derivation was published in Reference 5. During the AFE program, the algorithm was tested, compared, and evaluated against other guidance algorithms in three and six degree-of-freedom computer-based simulations.[6] The HYPAS guidance algorithm was selected for the space flight test, and development of the flight code was on schedule until the AFE program was cancelled.

The HYPAS algorithm has been used in numerous human and robotic exploration mission studies performed at NASA over the last several years. These studies involved developing nominal and dispersed trajectory simulation results for aerocapture at Earth and Mars, for a wide range of vehicle L/D, ballistic coefficients, entry conditions, and target orbits. This work has provided the opportunity to gain a deep understanding of how the guidance algorithm performs in a variety of situations. Modifications have been made as necessary to improve performance and robustness. These modifications include maintaining the equilibrium glide drag reference into the exit phase, correction for measured L/D (from sensed acceleration vector) in the commanded bank equation, smoothing the transition between the capture and exit phases, and calculation of bank commands before atmosphere entry based on the estimated position in the entry corridor.

TPC

The TPC algorithm also uses bank control to guide a lifting vehicle to a desired apoapsis and inclination or plane. TPC is based on a calculus of variations approach and is analogous to the Apollo Earth-entry guidance, but with different boundary conditions. TPC is a feedback guidance that uses sensitivities of the exit condition to changes in the state and control to determine the control at any point along the trajectory.

The sensitivities are generated from a reference trajectory that is determined off-line prior to flight. The guidance does not attempt to follow the reference trajectory or any other particular trajectory. The reference trajectory is simply used to generate the sensitivity coefficients. This reference trajectory is run open-loop with a reduced lift coefficient. Reducing the lift coefficient gives the same effect in apoapsis as a nonvertical bank angle without the out-of-plane effects. Any valid trajectory can be used as a reference, but usually one with a constant bank angle that meets the target condition is chosen.

The TPC algorithm has been studied extensively for several proposed aerocapture missions. It was originally developed for the Mars Surveyor Program 2001 Orbiter. The French space agency (CNES) ranked it best among several algorithms proposed for the "Premier" sample return mission before aerocapture was dropped from the mission. The analogous lander guidance was chosen for the Apollo Earth-entry guidance because it needed very little onboard computer resources, it was very accurate and it degraded gracefully under extreme conditions.

TPC has an in-plane component, which targets the velocity increment (ΔV) required to achieve a desired orbit after the atmospheric pass is complete, and an out-of-plane component, which targets inclination or wedge angle. The out-of-plane logic is structurally similar to the HYPAS out-of-plane logic and relies on a deadband to trigger bank angle reversals.

Derivation of the gains used by the TPC algorithm is presented in Reference 7. A reference trajectory is generated off-line. The adjoint equations to the equations of motion are then integrated from the atmospheric exit backwards to the atmospheric entry. The adjoint variables (sensitivity coefficients) are used to create gains on drag acceleration, altitude-rate, and velocity. A data point reduction algorithm is then applied to the gains to reduce the size of the onboard data required. Note that this process is completely automated after the reference trajectory is chosen.

On board, the guidance basically does a table lookup and multiplies the stored gains by the appropriate states. This determines a bank angle that will drive the final ΔV to be near the reference ΔV. The entire trajectory is flown the same way and is treated as a single continuous passage. For cases with an extremely low apoapsis, an override command is applied to help raise the apoapsis.

Density estimation of the atmosphere is also performed on board. The TPC density estimation uses a least squares fit of the most recent acceleration measurements in a manner which is theoretically very similar to that used by HYPAS, though the two implementations are unique.

RECENT TESTING OF THE HYPAS AND TPC ALGORITHMS

During 1998, the HYPAS and TPC algorithms were investigated for use on the Mars Surveyor Program 2001 mission. The algorithms were tested in computer-based simulation environments at JSC and LaRC and found to perform well under nominal and dispersed conditions. The results from that work are published in Reference 7 and Reference 8. The algorithms underwent extensive analysis and testing for the Mars Sample Return Orbiter and the Mars Premier Mission studies that were jointly conducted with the French Space Agency, CNES. Again, the algorithms were found to perform very well in computer based simulations. The results from that work are published in References 9, 10, and 11.

Although the performance of these algorithms has been studied extensively for Earth and Mars, there has been no previous analysis of their capabilities and performance for aerocapture at other destinations. The remainder of this paper describes a recent assessment for aerocapture at Titan.

PERFORMANCE FOR TITAN AEROCAPTURE

AEROCAPTURE ANALYSIS PROCESS

The aerocapture guidance analysis process involves several steps. First, the theoretical entry flight path angle corridor is determined. This is used to establish the nominal target entry conditions and sets the bounds on the entry delivery errors for the given vehicle's capabilities. Next, the theoretical optimum aerocapture performance is determined. This provides a benchmark for comparison with the performance obtained with the guidance algorithms. A nominal guided aerocapture trajectory is then developed. This involves an iterative process of adjusting guidance algorithm parameters and testing against variations in entry flight path angle, aerodynamic coefficients, and atmosphere density to get the best performance from the algorithm. Finally, Monte Carlo simulations are performed to evaluate the guidance performance under expected flight conditions.

A computer program that simulates atmospheric flight about a central body through three-dimensional space, plus rotation about the vehicle's velocity vector (bank angle) was used for this analysis. The central body, Titan, was assumed to have a spherical shape of radius 2575 km, and an inverse square gravity field, with gravitational parameter of 9142 km^3/s^2. Aerodynamic coefficients were assumed to be constant. The simulation uses an engineering model of the Titan atmosphere, which provides the capability to vary the density between minimum and maximum density profiles, as well as superimpose random density perturbations to simulate flight through a realistic atmosphere.[12] Further details of the trajectory simulation used for this analysis are explained in Reference 3.

ENTRY CORRIDOR

The first step in developing a guided aerocapture trajectory is to determine the theoretical entry flight path angle corridor. The steep side of the corridor is defined to be the flight path angle at which the vehicle just reaches the apoapsis target with full lift up for the duration of the atmospheric flight. The shallow side of the corridor is defined to be the flight path angle at which the vehicle just stays below the target apoapsis with full lift down for the entire duration of the trajectory. The theoretical entry flight path angle corridor for the reference mission was determined using

Table 1. Entry Corridors for Titan Aerocapture

Atmosphere Profile	Steep Side (deg)	Shallow Side (deg)	Width (deg)	Middle (deg)
Nominal	-37.62	-34.15	3.46	-35.88
Minimum	-38.37	-35.28	3.09	-36.82
Maximum	-36.78	-32.91	3.87	-34.85
Combined	-36.78	-35.28	1.50	-36.03

the nominal, minimum, and maximum density atmosphere profiles and is shown in Table 1. Taking into account the total uncertainty in the atmosphere profiles, the entry flight path angle corridor width is 1.5 deg. This is sufficient for the expected entry delivery errors.[13] The nominal entry flight path angle was chosen to be the middle of the combined corridors, which is at -36.03 deg.

THEORETICAL OPTIMUM PERFORMANCE

With the theoretical entry corridors defined, the theoretical optimum performance can be determined. The theoretical optimum guidance exactly achieves the target apoapsis while maximizing the periapsis altitude (thus minimizing the post-aerocapture ΔV required) for all flight path angles within the theoretical corridor. This is achieved by entering the atmosphere with full lift vector up, and then switching to full lift vector down at the correct instant to reach the target apoapsis altitude. The point at which the switch from lift vector up to lift vector down occurs is a function of the entry flight path angle. For entry at the shallow side of the corridor, the switch to lift vector down occurs immediately at entry. For entry at the steep side of the corridor, the switch to lift vector down is not performed. The theoretical optimum guidance performance is easy to determine, but it is not practical to achieve in real flight. The usefulness of knowing the optimum performance is to evaluate how well practical guidance algorithms perform compared to the theoretical optimum.

The theoretical optimum performance was determined for the reference mission and the nominal atmosphere profile for entry flight path angles across the theoretical corridor. The theoretical minimum ΔV required to circularize the orbit after aerocapture ranges from 224 m/s at the steep side of the entry corridor to 149 m/s at the shallow side of the corridor. For entry at the middle of the corridor (-36 deg), the periapsis altitude achieved is 290 km, which requires an ideal ΔV of 152 m/s after aerocapture to circularize the orbit at 1700 km altitude.

Once the theoretical corridor and optimum guidance performance were defined, the performance of the HYPAS and TPC algorithms were evaluated.

HYPAS PERFORMANCE

The HYPAS guidance algorithm parameters were adjusted for the reference aerocapture mission, and a nominal aerocapture trajectory developed. Figure 1 shows the results from the nominal trajectory. Trajectory simulations with the HYPAS guidance algorithm were run with the entry flight path angle varied across the theoretical corridor for the nominal atmosphere profile. The results can be seen in Figure 2, which shows apoapsis altitude achieved and post-

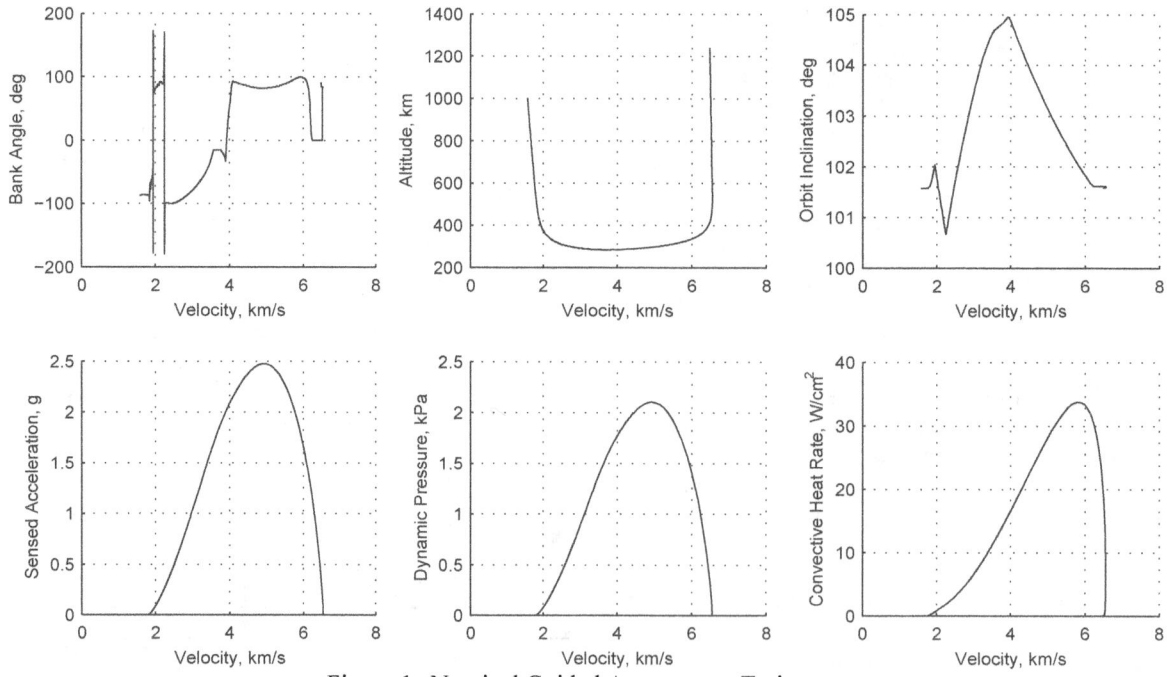

Figure 1. Nominal Guided Aerocapture Trajectory

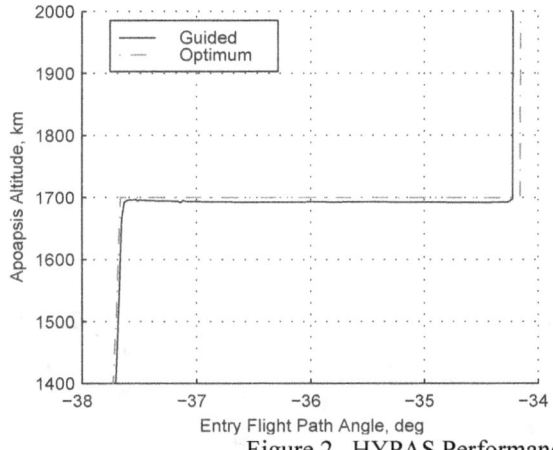 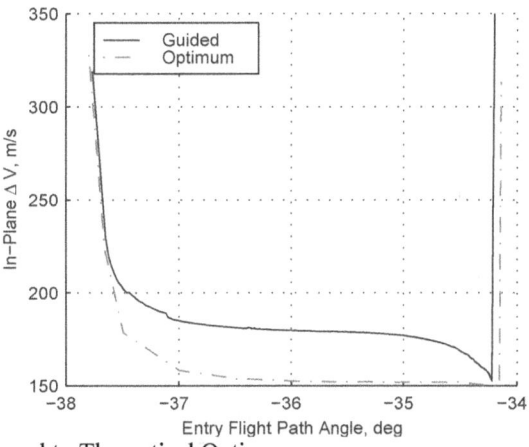

Figure 2. HYPAS Performance Compared to Theoretical Optimum

aerocapture ΔV required versus entry flight path angle for the HYPAS and theoretical optimal guidance. The guided entry corridor was found to extend from -37.6 to -34.2 deg, which covers 98 percent of the theoretical corridor. The post-aerocapture ΔV required with the HYPAS guidance at the middle of the corridor is 180 m/s, which is only 18 percent higher than the theoretical minimum.

Trajectory simulations with the HYPAS guidance algorithm were also run with the entry flight path angle varied across the theoretical corridor for the minimum and maximum atmosphere profiles to determine the sensitivity to variation in the atmosphere density. Initially, it was found that the algorithm's accuracy degraded significantly with the minimum and maximum density profiles. The reason for this is that the HYPAS algorithm assumes that the atmosphere density varies exponentially with altitude to predict exit conditions. The density scale height used in this model is a constant that is input prior to flight. This approach works well for missions where the atmosphere scale height does not vary significantly. However, the scale height of the Titan atmosphere does vary significantly between the minimum and maximum expected density profiles. The scale height for the nominal profile is approximately 50 km; for the minimum and maximum profiles, the scale height is approximately 42.5 and 53 km, respectively. This is a variation of –15 to +6 percent from the nominal and was found to be the source of the error in apoapsis altitude achieved.

The variation in atmosphere scale height can be accounted for using on-board estimation of the atmosphere scale height from navigated altitude and sensed acceleration.[14] A least squares filter was added to the HYPAS logic to do this. The results with and without the scale height estimation are shown in Figure 3. As can be seen, with the scale height estimation incorporated, the HYPAS guidance is insensitive to variation in the atmosphere density profile.

An assessment of the HYPAS algorithm performance when subject to uncertainties in aerodynamic coefficients was made. Trajectory simulations were run with the entry flight path angle varied across the theoretical corridor for variations of ±10 percent in the lift and drag coefficients. The results are shown in Figure 4. As can be seen, the algorithm is insensitive to the variation in aerodynamic coefficients, which is due to the automatic adjustment in the bank angle command the guidance makes using the sensed acceleration.

Monte Carlo simulations, which combine random variations in entry position and velocity vectors, aerodynamic coefficients, vehicle mass, mean atmosphere density profile, plus random atmosphere density perturbations, were performed to complete the assessment of the HYPAS algorithm's performance for the reference Titan aerocapture mission. Figure 5 shows a sample of the results for a 2000 case Monte Carlo with a nominal entry velocity of 6.5 km/s. Several other sets of Monte Carlo simulations were run with varying sets of assumptions. The results of those simulations can be found in Reference 3. The guidance algorithm was found to perform very well for the reference Titan aerocapture mission.

TPC PERFORMANCE

A nominal reference trajectory for the titan aerocapture was run and used to generate gains for the TPC guidance algorithm. The bank angle profile of the nominal trajectory is shown in Figure 6. The reference trajectory used a constant bank angle of about 62 degrees.

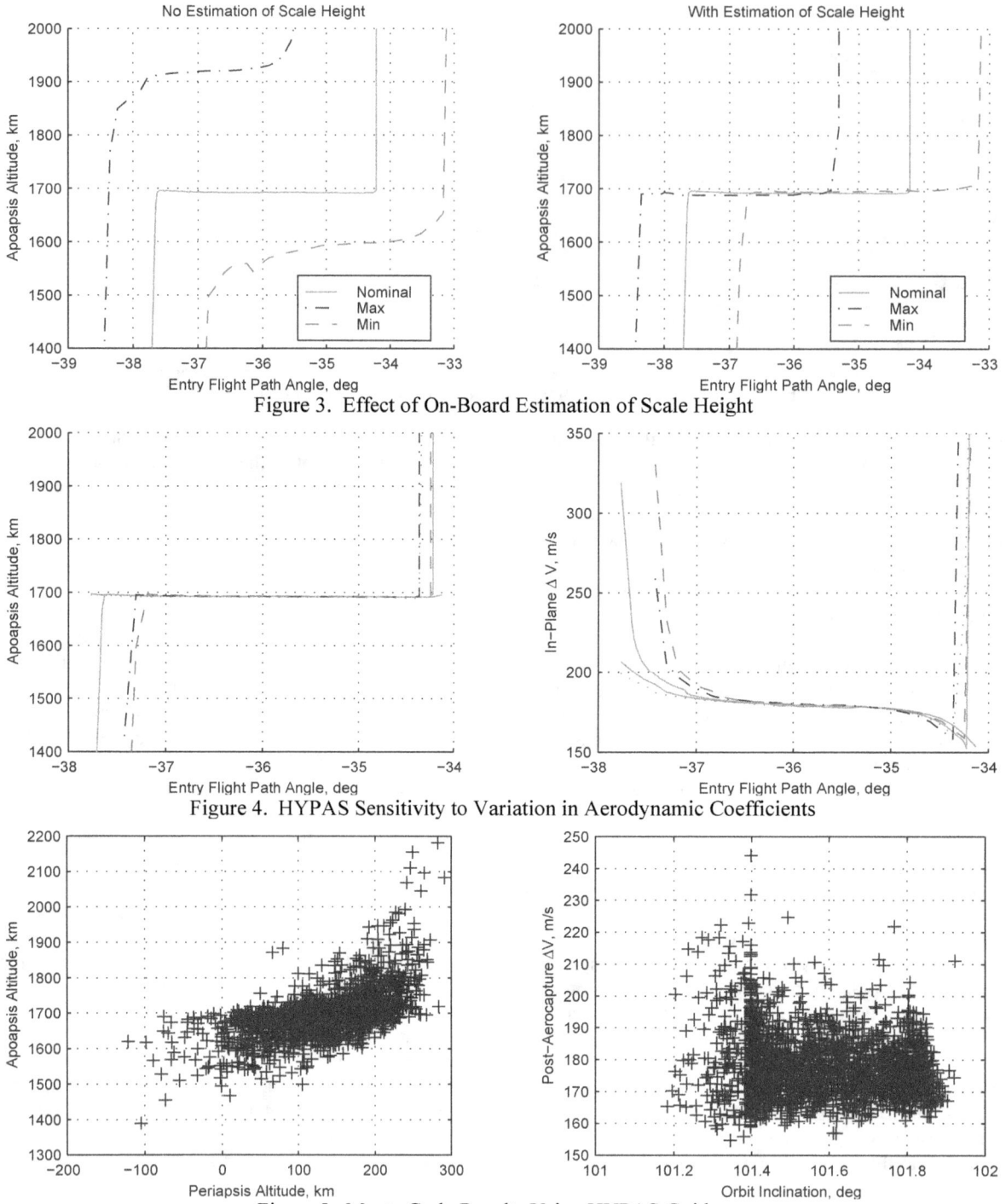

Figure 3. Effect of On-Board Estimation of Scale Height

Figure 4. HYPAS Sensitivity to Variation in Aerodynamic Coefficients

Figure 5. Monte Carlo Results Using HYPAS Guidance

For these initial test runs, a simple "funnel" deadband is used. The deadband is linear with velocity until a predefined velocity trigger (in this case 3000 m/s) at which point, the deadband switches to a constant width. The inclination history is also shown in Figure 6.

Monte Carlo simulations, as for the HYPAS algorithm, were performed to complete the assessment of the TPC algorithm. Figure 7 shows the resulting periapsis and apoapsis when the vehicle left the atmosphere for 2000 cases. None of these cases failed to capture, either high (skip out) or low (crash). A strong trend is obvious that the cases with high periapsis also have high apoapsis. The primary factor affecting the high cases seems to be initial flight path angle. All of the cases that ended with high apoapsis had a shallow entry flight path angle, though many cases with a shallow entry angle ended with apoapsis near the nominal.

Figure 7 also shows inclination and ΔV to circularize the orbit at 1700 km. The deadband on inclination has performed well. All cases are within 0.5 deg of the target inclination of 101.6 deg. Most of the cases have a ΔV near 180 m/s, though a significant minority are clustered near 190 m/s. 99 percent of the cases have a final ΔV less than 201 m/s.

SUMMARY AND CONCLUSIONS

An evaluation of two different guidance algorithms, the HYPAS and the TPC, has been completed as part of an assessment of the feasibility of aerocapture at Titan. This evaluation included determination of the theoretical entry corridor and comparison of the

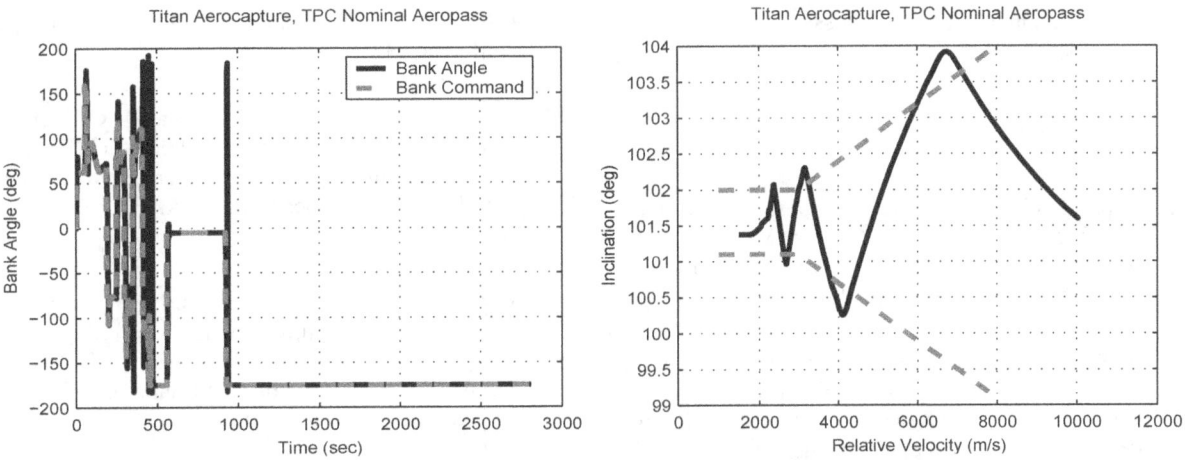

Figure 6. Nominal Bank and Inclination Profile Using TPC Guidance

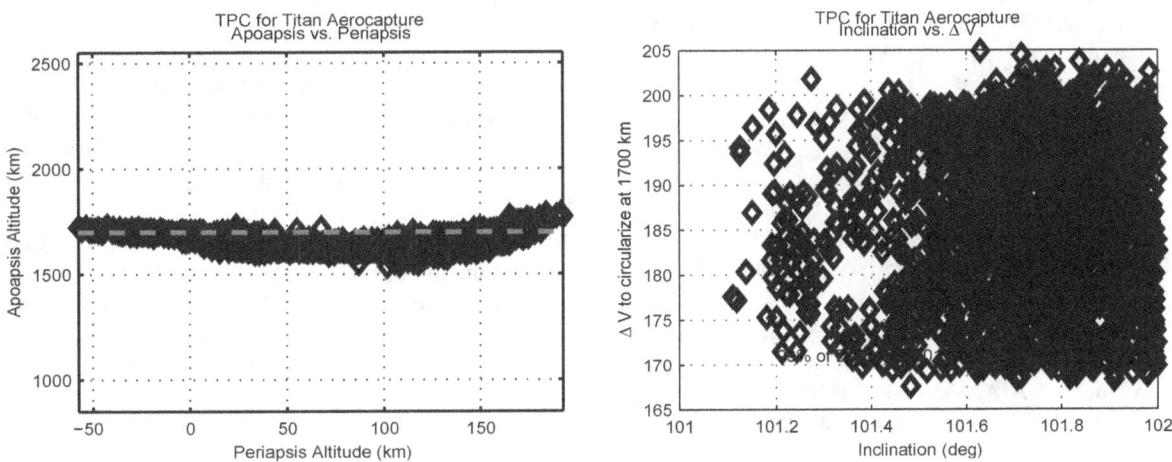

Figure 7. Monte Carlo Results Using TPC Guidance

guidance algorithms' performance to the theoretical optimum. Guidance responses to off-nominal conditions, including Monte Carlo simulations, were also investigated.

The HYPAS and TPC algorithms have been previously studied for aerocapture at Earth and Mars, and they were used at Titan without any significant modification. The only difference is that on-board estimation of the atmosphere density scale height was included to improve guidance accuracy. This is because there is a large variation in current estimates of the scale height of Titan's atmosphere.

Both aerocapture guidance algorithms were shown to perform well for the proposed Titan Explorer mission. No major technical issues are expected in implementing a guidance algorithm for this reference mission. Therefore the Titan aerocapture mission is feasible from a guidance algorithm perspective.

REFERENCES

1. M.K. Lockwood, "Titan Aerocapture Systems Analysis," AIAA-2003-4799, AIAA/ASME/SAE/ASEE Joint Propulsion Conference, Huntsville, AL, July 2003.

2. R. Bailey, J. Hall, T. Spilker, "Titan Aerocapture Mission and Spacecraft Design Overview," AIAA-2003-4800, AIAA/ASME/SAE/ASEE Joint Propulsion Conference, Huntsville, AL, July 2003.

3. D. Way, R.W. Powell, J. Masciarelli, B. Starr, K.T. Edquist, "Aerocapture Simulation and Performance for the Titan Explorer Mission," AIAA-2003-4951, AIAA/ASME/SAE/ASEE Joint Propulsion Conference, Huntsville, AL, July 2003.

4. B. Bragg, C. Cerimele, R. Delventhal, J. Gamble, O. Hill, R. Kincade, D. Lee, W. Long, R. McHenry, G. McSwain, K. Nagy, M. Richardson, R. Ried, B. Roberts, C. Scott, D. Smith, "A Design Study for an Aeroassist Flight Experiment," NASA Johnson Space Center, JSC-20593, June 1985.

5. C. Cerimele, J. Gamble, "A Simplified Guidance Algorithm for Lifting Aeroassist Orbital Transfer Vehicles," AIAA-85-0348, AIAA 23rd Aerospace Sciences Meeting, Reno, Nevada, January 1985.

6. T. Snook, R. McHenry, "Monte Carlo Evaluation of Aerobraking Guidance Algorithms," NASA Johnson Space Center, JSC-22432, January 1987.

7. T. Ro, E. Queen, "Mars Aerocapture Terminal Point Guidance and Control," AIAA-98-4571, AIAA Atmospheric Flight Mechanics Conference, Boston, MA, August 1998.

8. L. Bryant, M. Tigges, D. Ives, "Analytic Drag Control for Precision Landing and Aerocapture," AIAA-98-4572, AIAA Atmospheric Flight Mechanics Conference, Boston, MA, August 1998.

9. J. Masciarelli, S. Rousseau, H. Fraysse, E. Perot, "An Analytic Aerocapture Guidance Algorithm for the Mars Sample Return Orbiter," AIAA-2000-4116, AIAA Atmospheric Flight Mechanics Conference, Denver, CO, August 2000.

10. E. Perot, H. Fraysse, S. Rousseau, J.C. Berges, "Comparison of an Analytical Predictor Corrector and a Terminal Point Controller for the Mars Sample Return Aerocapture," AAAF 14-67 26-29 March 2001.

11. S. Rousseau, E. Perot, C. Graves, J. Masciarelli, E. Queen, "Aerocapture Guidance Algorithm Comparison Campaign," AIAA-2002-4822, AIAA/AAS Astrodynamics Specialist Conference, Monterey, CA, August 2002.

12. C.G. Justus, A. Duvall, and D.L. Johnson, "Engineering-Level Model Atmospheres For Titan And Neptune," AIAA-2003-4803, AIAA/ASME/SAE/ASEE Joint Propulsion Conference, Huntsville, AL, July 2003.

13. R. Haw, "Titan Approach Navigation for the Titan Aerocapture Orbiter," AIAA-2003-4802, AIAA/ASME/SAE/ASEE Joint Propulsion Conference, Huntsville, AL, July 2003.

14. E. Perot, S. Rousseau, "Importance of an On-Board Estimation of the Density Scale Height for Various Aerocapture Guidance Algorithms," AIAA-2002-4734, AIAA/AAS Astrodynamics Specialist Conference, Monterey, CA, August 2002.

AEROCAPTURE SIMULATION AND PERFORMANCE FOR THE TITAN EXPLORER MISSION

David W. Way
Richard W. Powell
Karl T. Edquist
James P. Masciarelli
Brett R. Starr

ABSTRACT

A systems study for a Titan aerocapture orbiter has been completed. The purpose of this study was to determine the feasibility and potential benefits of using aerocapture technologies for this destination. The Titan Explorer design reference mission is a follow-on to the Cassini/Huygens exploration of the Saturnian system that consists of both a lander and an orbiter. The orbiter uses aerocapture, a form of aeroassist, to replace an expensive orbit insertion maneuver with a single guided pass through the atmosphere. Key environmental assumptions addressed in this study include: the uncertainty in atmospheric density and high frequency atmospheric perturbations, approach navigation delivery errors, and vehicle aerodynamic uncertainty. The robustness of the system is evaluated through a Monte Carlo simulation. The Program to Optimize Simulated Trajectories is the basis for the simulation, though several Titan specific models were developed and implemented including: approach navigation, Titan atmosphere, hypersonic aeroshell aerodynamics, and aerocapture guidance. A navigation analysis identified the Saturn/Titan ephemeris error as major contributor to the delivery error. The Monte Carlo analysis verifies that a high-heritage, low L/D, aeroshell provides sufficient performance at a 6.5 km/s entry velocity using the Hybrid Predictor-corrector Aerocapture Scheme guidance. The current mission design demonstrates 3-sigma success without additional margin, assuming current ephemeris errors, and is therefore not dependent on the success of the Cassini/Huygens mission. However, additional margin above 3-sigma is expected along with the reduced ephemeris errors in the event of a successful Cassini mission.

NOMENCLATURE

BOC	Beginning of Cassini
c.g.	Center of Gravity
CAD	Computer Aided Design
CFD	Computational Fluid Dynamics
DoF	Degree of Freedom
EDL	Entry, Descent, and Landing
EOC	End of Cassini
GRAM	Global Reference Atmospheric Model
HYPAS	Hybrid Predictor-corrector Aerocapture Scheme
IRIS	Infrared Interferometer Spectrometer
L/D	Lift to Drag Ratio
LAURA	Langley Aerodynamic Upwind Relaxation Algorithm
MER	Mars Exploration Rover
MGS	Mars Global Surveyor
MPF	Mars Pathfinder
MRO	Mars Reconnaissance Orbiter
POST	Program to Optimize Simulated Trajectories
SEP	Solar Electric Propulsion
UVS	Ultraviolet Spectrometer
ΔDOR	Delta Differential One-way Ranging
ΔV	Velocity Addition
σ	Standard Deviation

BACKGROUND

AEROCAPTURE DESCRIPTION

Aerocapture, a form of aeroassist, is a propellant-less alternative to the currently requisite all-propulsive planetary capture. Using drag to decelerate the vehicle, aerocapture replaces the expensive orbit insertion maneuver with a single guided pass through the atmosphere. To date, aerocapture has not been demonstrated in flight.

In contrast, aerobraking uses many passes through the atmosphere to reduce the period of an elliptical orbit. This reduces, but does not eliminate, the propulsive capture requirement. Aerobraking has been used successfully in the Martian atmosphere by Mars Global Surveyor (MGS), and Mars Odyssey, and is planned for Mars Reconnaissance Orbiter (MRO).

A nominal drag profile associated with the aerocapture pass is designed to remove all of the hyperbolic excess velocity and enough additional orbital energy to place the spacecraft in an elliptical orbit with the desired apoapsis. Because of the larger energy requirements, aerocapture occurs at altitudes much lower than aerobraking. A guidance system is used to target the desired exit conditions by reacting to changes in the atmosphere. Bank angle modulation is used to control the rate of ascent/descent, which indirectly affects the drag. The flight path angles required to fly full lift-up and full lift-down form a theoretical entry corridor.

Figure 1 diagrams the sequence of aerocapture events. At the first apoapsis after the aerocapture pass, a small propulsive maneuver must be completed to raise the periapsis to the desired altitude. The periapsis must be raised during the first orbit in order to prevent the vehicle from re-entering the atmosphere a second time. Another small propulsive burn is typically performed at periapsis to clean-up any residuals in the desired science orbit apoapsis.

TITAN EXPLORER MISSION

The Titan Explorer design reference mission is a follow-on to the Cassini/Huygens exploration of the Saturnian system that consists of both a lander and an orbiter.[1] Both spacecraft are launched together on a single Delta IV-class launch vehicle in 2010. Figure 2 shows a

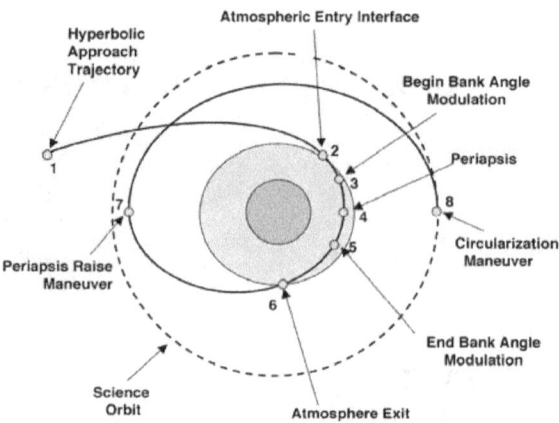

Figure 1: Aerocapture into Circular Orbit

Computer Aided Design (CAD) model of the stack packaged in a 4 m launch fairing. A Solar Electric Propulsion (SEP) module and a single Venus gravity assist provide a 6.25 year interplanetary cruise to the Saturn system.[2]

Both the orbiter and the lander are initially targeted for a direct entry to Titan. Thirty days prior to arrival, the orbiter releases the lander and executes a divert maneuver to the desired aerocapture approach trajectory.

The orbiter provides a telecom link for the lander during Entry, Descent, and Landing (EDL) then completes an aerocapture to the desired science orbit (a near-polar 1700 km circular orbit). Following aerocapture, the heatshield and backshell are jettisoned, and the orbiter begins a three-year science mission.

STUDY GOALS

A systems study for a Titan aerocapture orbiter has been completed as part of the NASA In-space Propulsion Program.[3] The purpose of this study was to determine the feasibility and potential benefits of using aerocapture technologies for this destination.[4,5] The products of this study are a reference mission, baseline systems definition, and technology requirements that may be used by scientists, systems engineers, technology developers, and mission managers in planning future missions. This study provides additional value over previous systems studies because of the higher fidelity of the analyses and environmental models that were employed.

Key environmental assumptions, central to successful aerocapture, are addressed in this study. These assumptions include the uncertainty in atmospheric density, high

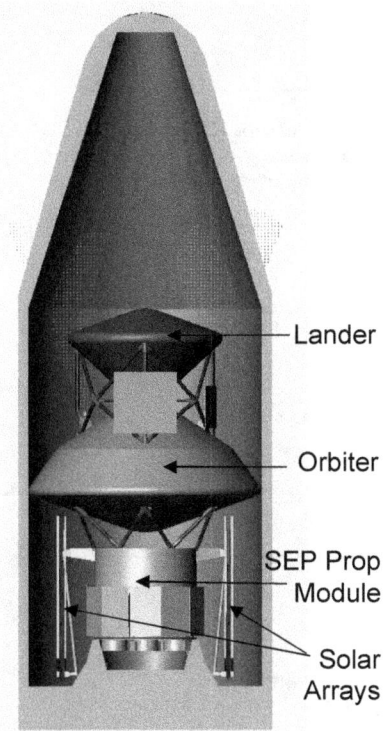

Figure 2: Launch Configuration

frequency atmospheric variability, approach navigation delivery errors, and vehicle aerodynamic uncertainty. Aerocapture risk is mitigated by quantifying the atmospheric uncertainty based on all available measurements, designing the vehicle to provide adequate aerodynamic control authority, developing a robust guidance system, and incorporating sufficient margins. The robustness of the system is evaluated through Monte Carlo simulation.

SIMULATION HERITAGE

The heritage of the Monte Carlo simulation used in this study is based upon previous Langley Research Center work on many diverse planetary missions that involve phases of atmospheric flight. These missions include aerobraking orbiters: MGS, Mars Odyssey, and MRO (scheduled for launch in 2005); direct lander entries: Mars Pathfinder (MPF), Genesis, Stardust, Mars 2001 Surveyor Lander (cancelled), Mars Exploration Rovers (MER), and Mars Science Laboratory (MSL) (planned for launch in 2009); and aerocapture proposals: Mars Surveyor 2001 Orbiter (cancelled) and Mars Premier Orbiter (aerocapture option not adopted). The current simulation leverages this experience in atmospheric flight and applies it to a new destination, Saturn's largest moon, Titan.

SIMULATION DEVELOPMENT

TRAJECTORY SIMULATION

To aid in the systems study activity, a high fidelity Monte Carlo trajectory simulation has been developed to simulate flight through the Titan atmosphere during aerocapture. This simulation provides data and statistics used to quantify mission success probabilities, evaluate candidate guidance algorithms, and provide the technical feedback required for mission and aeroshell design (aerodynamic loads, maximum heat rate, integrated heat loads, orbit circularization fuel, etc.).

The Program to Optimize Simulated Trajectories (POST) is the basis for this simulation.[6] However, several Titan specific models were developed and implemented to support this work. These models include: approach navigation, Titan atmosphere, hypersonic aeroshell aerodynamics, and aerocapture guidance. These models are discussed in more detail.

ATMOSPHERE

An engineering-level atmosphere model, denoted Titan-GRAM, was developed for this study.[7] Titan-GRAM is similar to and based upon the Mars Global Reference Atmosphere Model, Mars-GRAM, which has been used for in the design and operations support for many Mars exploration projects.

Titan-GRAM atmospheric density predictions are based on minimum, nominal, and maximum density vs. altitude profiles predicted by Yelle et al.[8] The Yelle models are based on observations from Voyager 1 radio science, Infrared Interferometer-Spectrometer (IRIS), and Ultraviolet Spectrometer (UVS). Shown in Figure 3, the Yelle density profiles include density variation due to latitude, season, and diurnal effects as well as measurement uncertainty.

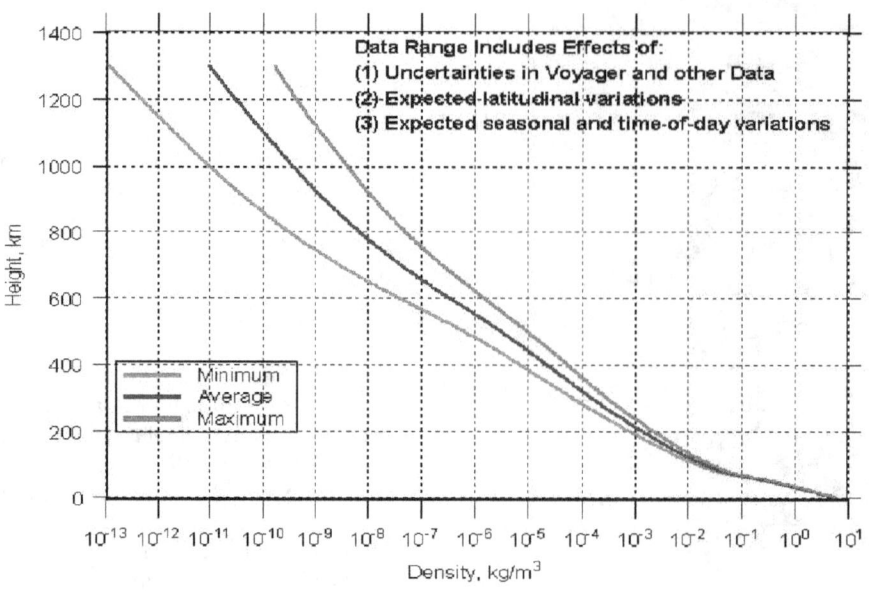

Figure 3: Yelle et al. Titan Atmospheric Density Profiles

Within Titan-GRAM, an atmospheric density control parameter, f_{minmax}, is used to linearly interpolate between the Yelle profiles. An f_{minmax} of 1.0 corresponds to the maximum expected density for a given altitude, while f_{minmax} of -1.0 corresponds to the minimum expected density. A sinusoidal variation of f_{minmax} with latitude was implemented to simulate latitudinal density gradients during an aerocapture pass.

Within the trajectory simulation, f_{minmax} is varied as a function of latitude to capture the expected latitudinal gradients. A perturbation model, based on gravity wave theory, is also included for use in the Monte Carlo analysis with a maximum perturbation (1σ) of 10% the mean density. Figure 4 shows a sample of perturbed density profiles generated by Titan-GRAM.

AERODYNAMICS

An aerodynamic model for the reference spacecraft has been developed using high-fidelity computations. The reference spacecraft has a 70 deg sphere-cone heatshield, similar to the Viking Mars Lander entry vehicle, and a bi-conic backshell. The configuration is shown in Figure 5.

Constant normal and axial force aerodynamic coefficients are used for the aerocapture pass simulation and are based on Langley Aerodynamic Upwind Relaxation Algorithm (LAURA) Computational Fluid Dynamics (CFD) results in the hypersonic regime. LAURA solves the viscous fluid dynamic equations on a structured grid with built-in adaptation.[9] Thermal and chemical non-equilibrium models are used to calculate the high-temperature flowfield behind the bow shock.

The high heritage, L/D = 0.25, aeroshell configuration provides 3.5 degrees of theoretical corridor width at a 6.5 km/sec entry velocity. A higher entry velocity of 10 km/s results in a 4.7 degree theoretical corridor.

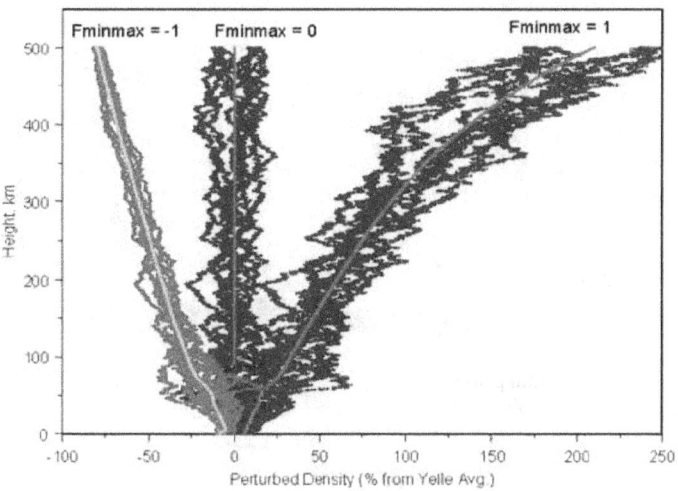

Figure 4: Sample Titan-GRAM Density Perturbations

GUIDANCE

A Hybrid Predictor-corrector Aerocapture Scheme (HYPAS) aerocapture guidance algorithm was developed and included in the simulation.[10] The HYPAS algorithm uses an analytic method, based on deceleration due to drag and altitude rate error feedback, to predict exit conditions and then adjust the bank angle command in order to achieve a target apoapsis altitude and orbit inclination at atmosphere exit.

The HYPAS guidance consists of two phases: the "capture phase", in which the guidance establishes pseudo-equilibrium glide conditions; and an "exit phase", in which exit conditions are predicted, assuming a constant altitude rate, and the lift vector is adjusted to null the error between predicted and target apoapsis. Figure 6 shows the guidance phases during an aerocapture pass.

Bank reversals maintain inclination error within desired limits. All reference values are computed and updated during flight. The HYPAS algorithm was adapted for use at Titan, and two sets of guidance initialization parameters were developed: one for the 6.5 km/s entry, and one for the 10.0 km/s entry. Monte Carlo trajectory simulations were run with this guidance to determine overall aerocapture performance.

A pseudo bank controller was developed to mimic the dynamics of a flight control system in a Three Degree-of-Freedom (3-DoF) simulation. These effects include a control system time lag and a finite system response, limited by a maximum angular acceleration and a maximum angular velocity. The bank angle controller analytically calculates the time required, and resulting angular travel necessary, to complete the maneuver to the commanded attitude. It has been found that including this type of controller in a 3-DoF simulation provides a good approximation to Six Degree-of-Freedom (6-DoF) dynamics.

Because the aerocapture spacecraft performs bank reversals to maintain inclination accuracy, and because these reversals could take as much as 15 seconds to complete, the trajectory simulation must model the effects of an attitude controller. These bank reversals force the spacecraft off of the optimum flight profile that

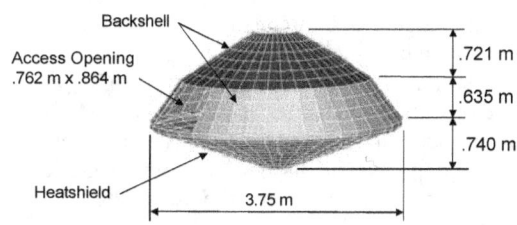

Figure 5: Aeroshell Configuration

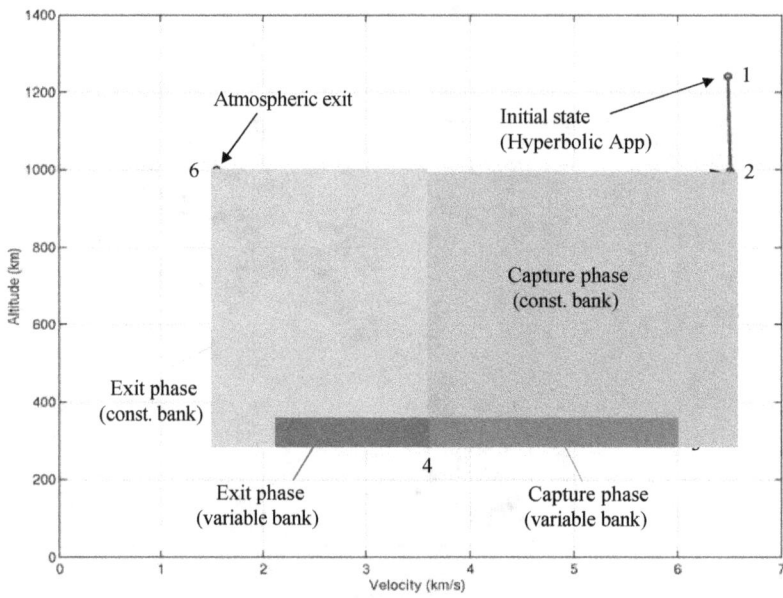

Figure 6: Phases of HYPAS Aerocapture Guidance

the guidance is trying to follow. Not including the error produced during this maneuver would result in overly optimistic conclusions regarding the vehicle's targeting ability and the required circularization ΔV.

NAVIGATION

Initial states were provided by a JPL navigation assessment that assumed post-Cassini ephemeris knowledge and the following data sources: two-way Doppler and ranging, ΔDOR, and optical navigation.[11] These assumptions resulted in a 3σ delivery flight path angle dispersion of ± 0.93 deg. Figure 7 shows the delivery footprint in the B-plane. The dashed line in this figure is a radius vector to the nominal aim-point.

The three dominant contributors to this delivery error were Saturn and Titan ephemeredes, maneuver execution error, and optical data measurement error. The current Cassini mission is expected to improve the ephemeris errors by a factor of six. However, this improved navigation is not guaranteed, but rather contingent upon the successful completion of the Cassini mission. Therefore, both Beginning of Cassini (BOC) and End of Cassini (EOC) states were evaluated. For the purposes of this study, it was assumed that the flight path angle dispersions would degrade by approximately 52% with the use of BOC states.

Since Titan is a moon of Saturn with an orbital period of approximately 16 days, the mission designer has a wide choice in approach velocities for any mission opportunity (Titan's velocity could either add or subtract from the nominal Saturn approach velocity). Intercepting Titan at different true anomalies easily tailors the entry velocity, with only small changes in the incoming hyperbolic approach trajectory. Entry velocities of 6.5 km/s and 10.0 km/s were considered. A 15% increase in flight path angle dispersions was levied on the higher entry velocity.

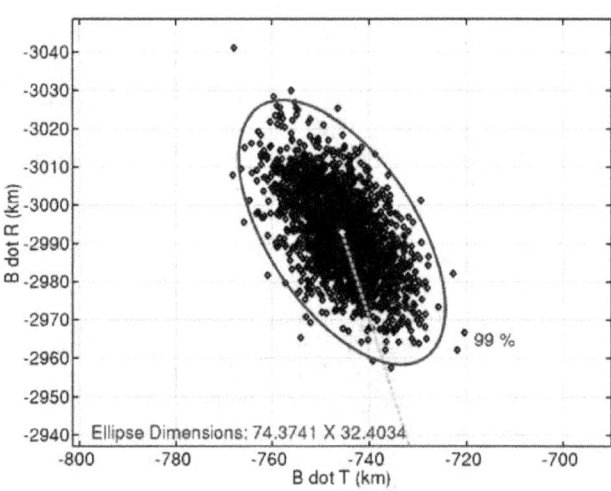

Figure 7: Delivery Error in the B-plane

The Table 1 summarizes the entry flight path angle dispersions assumed in this study along with the theoretical corridor for an L/D = 0.25. Further navigational assessment is required to validate these assumptions.

Table 1: Entry Flight Path Angle Uncertainties

Entry Velocity	EOC Ephemeris	BOC Ephemeris	Theoretical Corridor
6.5 km/s	± 0.93 deg	± 1.42 deg	3.5 deg
10 km/s	± 1.07 deg	± 1.63 deg	4.7 deg

RESULTS

MONTE CARLO ANALYSIS

System performance, risk, and robustness are measured by generating statistics from Monte Carlo simulations of the Titan aerocapture. Many (generally 2000) individual aerocapture trajectories are simulated with random perturbations applied to initial entry conditions, vehicle aerodynamics, vehicle mass properties, and Titan atmospheric conditions. This flight simulation is composed of three main parts: a POST2 trajectory simulation, which integrates all of the models discussed above; an executive Monte Carlo script, which coordinates the generation and execution of 16 parallel simulations; and various supporting scripts for sampling random distributions, compiling and

Table 2: Monte Carlo Uncertainties

Category	Variable	Nominal	± 3σ or min/max	Distribution
Initial Conditions				
	x- position	603.3 km	From covariance	Correlated
	y- position	-390.8 km	From covariance	Correlated
	z- position	3502 km	From covariance	Correlated
	x- velocity	-3.363 km/s	From covariance	Correlated
	y- velocity	-4.123 km/s	From covariance	Correlated
	z- velocity	-3.734 km/s	From covariance	Correlated
Atmosphere				
	Perturbation seed	1	1/9999	Uniform (integer)
	Fminmax bias	0	-0.53/+0.53	Uniform
Aerodynamics				
	Trim angle-of-attack	-16 deg	± 2.0 deg	Normal
	C_A (axial force)	1.48	± 3%	Normal
	C_N (normal force)	-0.05	± 5%	Normal
Mass Properties				
	Axial c.g.(Z_{cg}/D)	0.1979	± 0.00848	Normal
	Radial c.g.(X_{cg}/D)	0.0231714	± 0.00184	Normal

formatting output data, evaluating metrics and statistics, and producing plots and figures. Table 2 lists the uncertainties used in this study.

BEGINNING OF CASSINI

The first scenario examined is for a navigation ephemeris uncertainty consistent with knowledge prior to the Cassini mission. This combination of large navigation uncertainty and low entry velocity, 6.5 km/s, results in the most challenging conditions. Statistics for apoapsis altitude and circularization ΔV are presented in Table 3.

Figure 8 shows the aerocapture corridor (flight path angle) as a function of f_{minmax} (density). The theoretical corridor is bounded by the full lift-down and full lift-up cases. The plus (+) symbols show the range of f_{minmax} during the active guidance portion of the aerocapture pass, due to latitudinal variation of f_{minmax}. The circles indicate the f_{minmax} at periapsis. The bias in the data towards the higher values of f_{minmax} is again due to the latitudinal variation of f_{minmax}, since the aerocapture pass occurs over northern latitudes. The target flight path angle was chosen to capture as many of the cases as possible into the theoretical corridor. The large spread in entry flight path angle, compared to the theoretical corridor, suggests small margins.

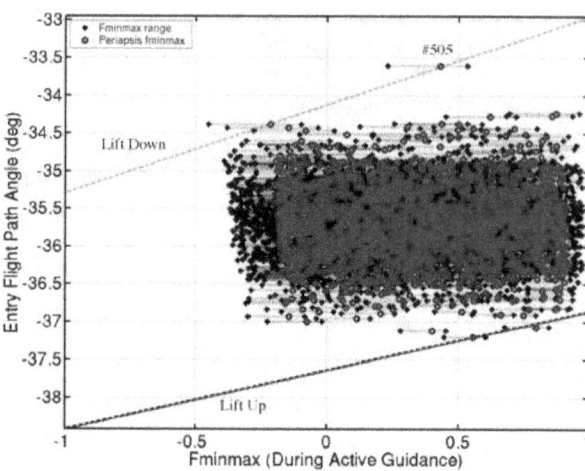

Figure 8: Entry Corridor, BOC 6.5 km/s

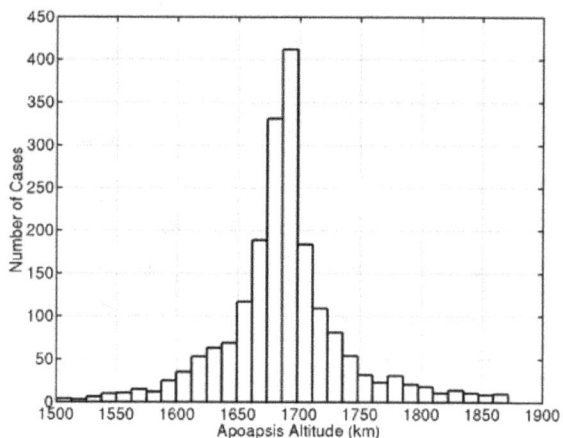

Figure 9: Apoapsis Altitude, BOC 6.5 km/s

Table 3 Performance Statistics, BOC 6.5 km/s

	Apoapsis altitude (km)	Circularization ΔV (m/s)
Minimum	1240.5	157.2
Maximum	2166.3	293.3
Mean	1691.3	179.6
1σ	± 63.9	± 11.1
3σ	± 191.8	± 33.2

Figure 9 shows a histogram of the final apoapsis altitude. The mean apoapsis of 1691.3 km, with a standard deviation of +/- 63.9 km, compares well with the 1700 km target altitude. Additionally, the 3σ range of +/- 191.8 km is within the target range of < 200 km, which indicates that the guidance can be tuned to capture a large percentage of the theoretical corridor. Only one case (#505) failed to capture. However, this case represents a 4.5-sigma case for entry flight path angle, which has a probability of occurrence of only 1 in nearly 15,000.

Figure 10 shows the histogram for the required ΔV. This ΔV includes both the periapsis raise maneuver and the final circularization burn. The 3-sigma (99.86 %-tile) value is 212.9 m/s.

END OF CASSINI

These Monte Carlo results are representative of navigation ephemeris uncertainty post-Cassini for a 6.5 km entry velocity. Statistics for

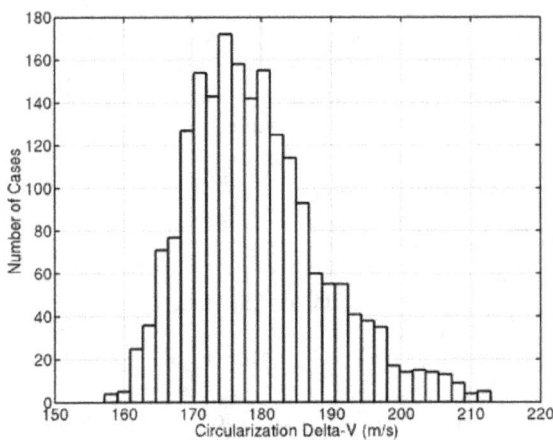

Figure 10: Circularization ΔV, BOC 6.5 km/s

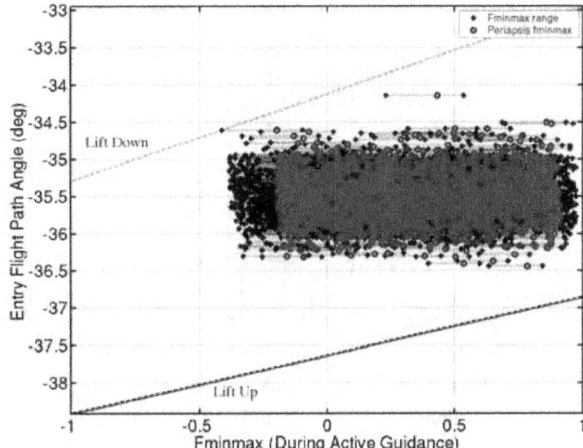

Figure 11: Entry Corridor, EOC 6.5 km/s

apoapsis altitude and circularization ΔV are presented in Table 4.

Figure 11 shows the aerocapture corridor (flight path angle) as a function of f_{minmax} (density). The effect of the improved (reduced) uncertainty expected from the Cassini mission is evident by the tighter grouping in flight path angle as compared to Figure 8. The size of the flight path angle dispersions, compared to the theoretic corridor, suggests increased margins for this scenario. Because of the arrival geometry, the aerocapture pass occurs over northern latitudes – entering over the northern pole and exiting near the equator. Therefore, the mean f_{minmax} (~0.3) is positive. The target flight path angle was chosen to bring this mean to the center of the theoretical corridor.

Table 4 Performance Statistics, EOC 6.5 km/s

	Apoapsis altitude (km)	Circularization ΔV (m/s)
Minimum	1327.8	156.0
Maximum	2196.6	252.8
Mean	1697.7	177.7
1σ	± 63.5	± 9.5
3σ	± 190.4	± 28.6

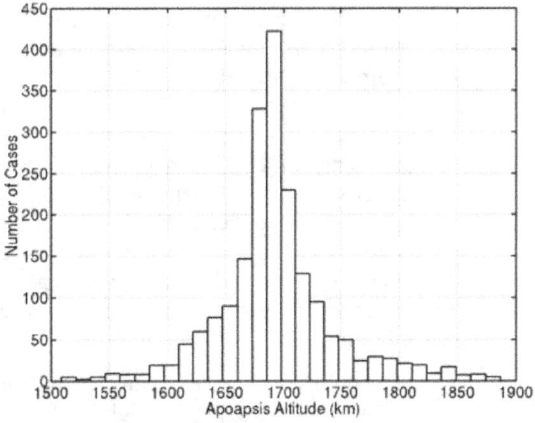

Figure 12: Apoapsis Altitude, EOC 6.5 km/s

Figure 12 shows a histogram of the final apoapsis altitude. The mean apoapsis is 1697.7 km, with a standard deviation of 63.5 km. The 3-sigma range of +/- 190.4 km is within the target range of < 200 km. All cases captured within 500 km of the target apoapsis.

Figure 13 shows the histogram and statistics for the required ΔV. The 3-sigma (99.86 %-tile) value is 206.3 m/s. The maximum ΔV was 252.8 m/s.

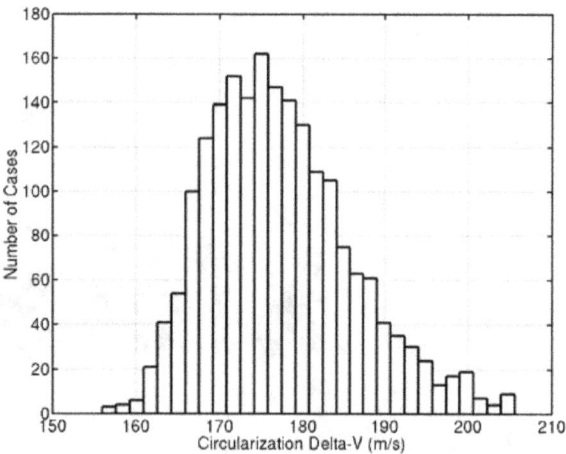

Figure 13: Circularization ΔV, EOC 6.5 km/s

CONCLUSIONS

1. The JPL navigation analysis identified the Saturn/Titan ephemeris error as major contributor the delivery error. The current mission design demonstrates 3-sigma success, without additional margin, assuming BOC ephemeris errors, and is therefore not dependent on the success of the Cassini/Huygens mission. However, additional margin above 3-sigma is expected along with the reduced EOC ephemeris errors in the event of a successful Cassini mission.

2. Uncertainty in the Titan atmospheric density, including high frequency perturbations, is the single largest unknown. To mitigate this risk, sufficient margin and conservatism are carried in the design of the entry conditions, aeroshell, and guidance system. While arrival during a particular season would reduce the expected density range, the full density range was used in the Monte Carlo analysis.

3. The Monte Carlo analysis verifies that a high-heritage, low L/D, aeroshell provides sufficient performance at a 6.5 km/s entry velocity. A mid L/D aeroshell technology development is not required. Additional aerocapture performance is also available at higher entry velocities, 10 km/s, but must be traded against increased Thermal Protection System requirements.

4. The Monte Carlo analysis demonstrates that the HYPAS guidance is robust and provides acceptable performance. Approximately 92% of the theoretical corridor is captured using this algorithm while requiring only slightly more than 200 m/s of on-orbit ΔV to achieve the target science orbit.

REFERENCES

1. Bailey, R., Hall, J., and Spilker, T., "Titan Aerocapture Mission and Spacecraft Design Overview," AIAA-2003-4800, Conference Proceedings of the 39th AIAA/ASME/SAE/ASEE Joint Propulsion Conference and Exhibit, Huntsville, AL, July 2003.

2. Noca, M., Bailey, R., and Dyke, R., "Titan Explorer Mission Trades from the Perspective of Aerocapture," AIAA-2003-4801, Conference Proceedings of the 39th AIAA/ASME/SAE/ASEE Joint Propulsion Conference and Exhibit, Huntsville, AL, July 2003.

3. James, B. and Munk, M., "Aerocapture Technology Development Within the NASA In-Space Propulsion Program," AIAA-2003-4654, Conference Proceedings of the 39th AIAA/ASME/SAE/ASEE Joint Propulsion Conference and Exhibit, Huntsville, AL, July 2003.

4. Hall, J., Noca, M., and Bailey, R., "Cost - Benefit Analysis of the Aerocapture Mission Set," AIAA-2003-4658, Conference Proceedings of the 39th AIAA/ASME/SAE/ASEE Joint Propulsion Conference and Exhibit, Huntsville, AL, July 2003.

5. Lockwood, M. K., "Titan Aerocapture Systems Analysis," AIAA-2003-4799, Conference Proceedings of the 39th AIAA/ASME/SAE/ASEE Joint Propulsion Conference and Exhibit, Huntsville, AL, July 2003.

6. Bauer, G. L., Cornick, D. E., and Stevenson, R., "Capabilities and Applications of the Program to Optimize Simulated Trajectories (POST)," NASA CR-2770, February 1977.

7. Justus, C. and Duvall, A., "Engineering-Level Model Atmospheres for Titan and Neptune," AIAA-2003-4803, Conference Proceedings of the 39th AIAA/ASME/SAE/ASEE Joint Propulsion Conference and Exhibit, Huntsville, AL, July 2003.

8. Yelle, R. V., Strobel, D. F., Lellouch, E., and Gautier, D., "Engineering Models for Titan's Atmosphere," in ESA report SP-1177, "Huygens Science, Payload and Mission," August 1997.

9. Cheatwood, F. M. and Gnoffo, P. A., "User's Manual for the Langley Aerodynamic Upwind Relaxation Algorithm (LAURA)," NASA TM-4674, April 1996.

10. Masciarelli, J. and Queen, E., "Guidance Algorithms for Aerocapture at Titan," AIAA-2003-4804, Conference Proceedings of the 39th AIAA/ASME/SAE/ASEE Joint Propulsion Conference and Exhibit, Huntsville, AL, July 2003.

11. Haw, R., "Approach Navigation for a Titan Aerocapture Orbiter," AIAA-2003-4802, Conference Proceedings of the 39th AIAA/ASME/SAE/ASEE Joint Propulsion Conference and Exhibit, Huntsville, AL, July 2003.

PRELIMINARY AEROTHERMODYNAMICS OF TITAN AEROCAPTURE AEROSHELL

Naruhisa Takashima
AMA Inc.
Hampton, VA 23666

Brian R. Hollis and E. Vincent Zoby
NASA Langley Research Center
Hampton, VA 23681

Kenneth Sutton
National Institute of Aerospace
Hampton, VA 23666

Joe Olejniczak
NASA Ames Research Center
Moffett Field, CA 94035

Michael J. Wright and Dinesh Prabhu
ELORET Corporation
Sunnyvale, CA 94087

ABSTRACT

Aeroheating environments for a Titan aerocapture mission were computed for the Titan Aerocapture Systems Analysis study funded through the In-Space Propulsion Program Office at NASA Marshall Space Flight Center. In this work, the convective heating environments for a candidate Titan probe (70 deg. sphere-cone geometry) are presented. The sensitivities of these environments to the computational grids, mass diffusion models, and reaction rates are examined in the context of axisymmetric flow using two different flow solvers. The lessons learned are applied to the forebody of the probe at an angle of attack of 16 deg. Results of computations from two different flow solvers are compared to reduce uncertainties in the predicted aerothermal environments. At the peak convective heating condition, the convective heating rate at the stagnation point is 28 W/cm^2 and the maximum heating rate, which occurs at the nose of the axisymmetric aeroshell is 46 W/cm^2. A smooth-wall transition criterion of Re_θ of 200 is used in determining the likelihood of flow transition on the forebody and shown to occur on the leeside. Turbulent flow calculations show that the level of heating on the leeside increases by 250%, and the location of the maximum heating point moves from the nose to the leeside shoulder. Finally, aerodynamic force coefficients extracted from the computed solutions are shown to be in good agreement with those assumed in the generation of flight trajectories. The three-dimensional aerothermal environments, along with the predicted radiative heating environments, are used in the selection and sizing of the Thermal Protection System for the forebody of the probe.

NOMENCLATURE

Alt	Altitude (km)
AoA	Angle of attack (deg)
CL	Lift coefficient
CD	Drag coefficient
D	Reference diameter (3.75 m)
L/D	Lift-to-Drag Ratio
Mach	Mach number
P	Pressure (Pa)
Re_θ	Momentum thickness Reynolds number.
ρ and Rho	Density (kg/m^3)
T	Translational temperature (K)
T_v	Vibrational temperature (K)
TPS	Thermal Protection System
V	Velocity (m/s)

INTRODUCTION

Titan, the largest moon of Saturn, is the only known moon in the solar system that has a fully developed atmosphere. The density of the atmosphere of Titan is actually greater than that of the Earth with a surface pressure 50% greater than that of Earth. The atmosphere is composed mainly of nitrogen, which accounts for approximately 94% by volume, with balance of argon and significant traces of hydrocarbon elements. Hydrocarbon elements are the building blocks for amino acids, a necessary ingredient for formation of life. For this reason, Titan is of significant interest to the scientific community for the understanding of early formation of life here on Earth. Voyager 1 was the first mission to Titan taking pictures of its dense atmosphere. Currently, the Cassini spacecraft with Huygens probe on board is en route to Titan. The joint NASA/ESA mission expects the spacecraft to arrive at Saturn on July 1st, 2004 and release its probe on December 25th, 2004.

As a potential follow-on mission to the Cassini-Huygens mission, a systems analysis study of an aerocapture mission to Titan was initiated by the NASA In-Space Propulsion Program Office at the NASA Marshall Space Flight Center (MSFC). Past studies have shown a vehicle mass saving of up to 66% can be realized using aerocapture technology compared to an all-propulsive mission to Titan. However, by design, these past analyses were of low fidelity and many assumptions were made to reach the conclusions. Hence, the goals for the present study include providing higher fidelity systems definition and sensitivities for Titan aerocapture mission that can be used by scientists, mission planners, technology planners, technologists and future mission managers; and to perform higher fidelity analyses for validating and updating previous assumptions. The system analysis study was performed by a multi-center NASA team with participation from NASA Ames Research Center (ARC), Jet Propulsion Laboratory, Johnson Space Center, Marshall Space Flight Center, and led by Langley Research Center (LaRC).[1]

This paper will present the results of the computations performed by LaRC and ARC to determine the aeroheating environments and will focus on the convective heating portion of these environments. A companion paper will present the results of the radiative heating analyses.[2] Prior to the work, extensive analyses, both numerical and experimental, were performed to determine the aeroheating environment of the Huygens probe.[3,4,5,6] Analyses have shown that significant radiative emission of mainly CN violet system can occur within the shock layer due to non-equilibrium condition, and the amount of the CN radiation is sensitive to the freestream gas composition. For the worst case condition, the level of radiative heat flux was determined to be twice that of the convective heating flux.[4]

TITAN AEROCAPTURE REFERENCE CONCEPT

The forebody design of the aeroshell used by the study is shown in Fig. 1. The design is based on the Mars Smart Lander (MSL) aeroshell design,[7] which has a 70 deg. sphere cone forebody and a bi-conic aftbody. For the Titan aeroshell, the MSL design was geometrically scaled to a diameter of 3.75 m, and the aftbody shape was altered to meet the packaging requirements. The overall length of the vehicle is 2.096 m, and the center of gravity is located such that an L/D of 0.25 is achieved at a trim angle of attack of 16 deg. The study defined six reference trajectories for an entry speed of 6.5 km/s with a ballistic coefficient of 90 kg/m^2. Trajectories were created for combinations of three atmospheric density profiles: minimum, nominal and maximum, with lift vector directions of up (undershoot) and down (overshoot). The three density profiles are shown in Fig. 2, which were obtained with Titan atmosphere model, TitanGRAM, developed by Justus and Duvall.[8] The reference trajectories were computed using the Program for Optimization of Simulated Trajectories (POST).[9]

For the present study, aerothermodynamic analyses were performed primarily for conditions along the minimum atmosphere lift-up trajectory. Engineering analyses showed that the maximum stagnation point convective heating rate on a sphere occurs along the minimum atmosphere lift-up trajectory. Fig. 3 shows the points along the reference trajectory that the computations were performed and Table 1 lists the freestream at each trajectory point. Based on equilibrium engineering analysis, cases 3 (t=253 s), 4 (t=281 s) and 5 (t=308 s) were predicted to be the peak convective heating, the peak dynamic pressure and the peak radiative heating, respectively.

There are many uncertainties in the composition of the Titan atmosphere.[8] The relative amounts of nitrogen, argon and methane will impact the overall heating rate on the vehicle. For radiative heating conservatism, the atmospheric composition was chosen as 95% N_2 and 5% CH_4 by volume along the entire reference trajectory.

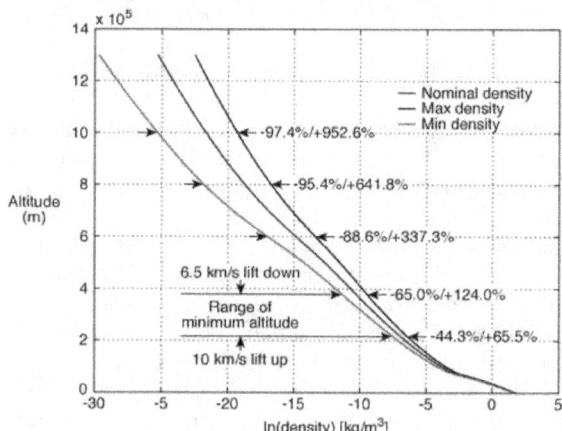

Fig. 2. Titan atmosphere density profiles.

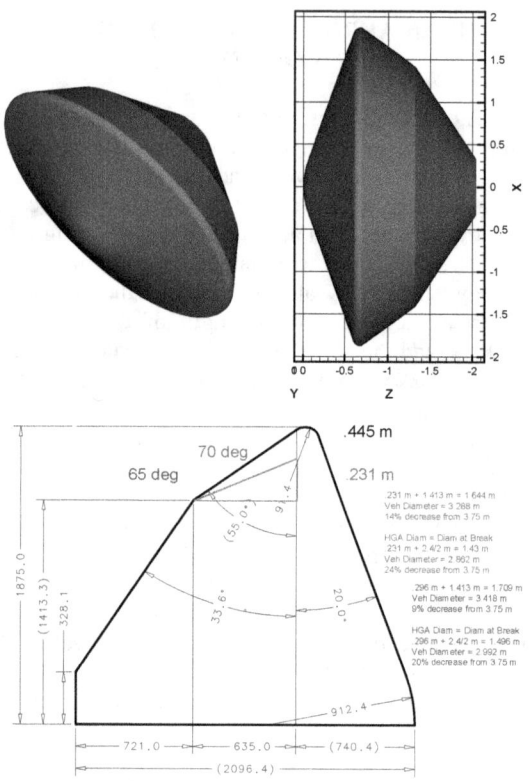

Fig. 1. Titan aeroshell design.

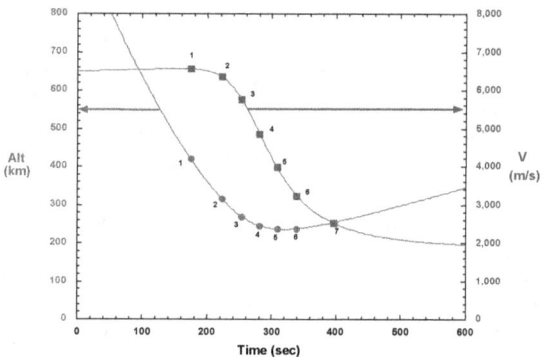

Fig. 3. Minimum atmosphere lift-up aeroheating reference trajectory.

Table 1. Freestream conditions along minimum atmosphere lift-up reference trajectory.

CASE	Time (sec)	Alt (km)	V (m/s)	AoA (deg)	rho (kg/m^3)	P (Pa)	T (K)	Mach
1	174	420	6558	16	4.697E-06	0.188	131.9	27.8
2	222	316	6348	16	5.070E-05	2.293	149.9	25.3
3	253	269	5761	16	1.491E-04	6.910	152.7	22.7
4	281	245	4859	16	2.665E-04	12.350	152.6	19.2
5	308	237	3978	16	3.305E-04	15.250	152.4	15.7
6	339	238	3239	16	3.212E-04	14.840	152.4	12.8
7	397	255	2521	16	2.138E-04	9.893	152.7	9.9

COMPUTATIONAL ANALYSES

As mentioned previously, the aeroheating environment computations were performed simultaneously during the course of the study by two teams from NASA Langley and Ames Research centers. Each team performed the calculations independently using different Computational Fluid Dynamics (CFD) programs. Once a set of calculations was made, the results were compared and analyzed by the two teams. If differences were observed, the causes were investigated and analyzed to the satisfaction of both teams. The purpose of having two totally independent parties calculating identical problems was to minimize the uncertainties in the results and gain greater confidence in the level of physical and numerical modeling in the codes.

During the course of the study, computational analyses were performed in two stages. During the first stage, computations were performed on the forebody at a zero degree angle of attack, i.e., axisymmetric flow, for the peak convective heating condition (Case 3) to perform sensitivity analyses with respect to transport/kinetic models. After the completion of the first stage, the second stage involved computing three-dimensional flow on the forebody for all seven trajectory points. The results from the second stage were delivered to the TPS team for TPS sizing purpose.

Sensitivity analyses were performed using two different CFD codes: Langley Aerothermodynamic Upwind Relaxation Algorithm (LAURA),[10] developed at LaRC, and the Data Parallel Line Relaxation (DPLR) program,[11] developed at ARC. Both CFD codes solve the three-dimensional Navier-Stokes equation using finite volume method with finite rate chemistry and thermal nonequilibrium. LAURA uses Roe's flux difference splitting with Yee's second-order symmetric total variation diminishing scheme to model the inviscid fluxes. Steady state solution is obtained using either point or line relaxation. The code supports MPI with multi-block structured grid to decrease computing time. DPLR uses third-order modified Steger-Warming flux vector splitting to calculate the inviscid fluxes and data-parallel line relaxation method to reduce the time to steady state convergence. Both codes use second order central differencing for the viscous fluxes and turbulent flow is modeled with the Baldwin-Lomax turbulence model. Both codes have been widely used in predicting aerothermal environments for several planetary probes.[7,10,12] Each code was executed by its respective center during the course of the study.

RESULTS

SENSITIVITY ANALYSES

To gain insight and reduce the uncertainties in the numerical results, sensitivity analysis on the diffusion model and chemical kinetic rates were made by performing calculations for the peak convective heating condition (Case 3) at zero degree of angle attack. For all the runs, the codes were executed with finite rate chemistry and thermal nonequilibrium using Park's two-temperature model.[13] For the nominal case, a total of 18 species was assumed to be present in the gas mixture: N, N^+, N_2, N_2^+, C, C^+, C_2, CH, CH_2, CH_3, CH_4, H, H^+, H_2, CN, CN^+, NH and e^-. "Super catalytic" surface boundary condition was chosen where the species fraction is fixed at freestream values of 95% N_2 and 5% CH_4 at the wall. Forcing the gas mixture to recombine to the freestream composition causes the chemical energy to be released back into the flow, which results in conservative wall heating rates. The wall temperature was assumed to be at radiative equilibrium with surface emissitivity of 0.90, and the boundary layer was assumed to be laminar.

Fig. 4. shows the grid and the temperature contour for the axisymmetric solution using LAURA. The grid for the solution was obtained using the built-in grid alignment capability of LAURA. The capability allows for grid clustering within the boundary layer and near the shock, as well as, adaptation of the outer boundary to the bow shock. The degree of adaptation is controlled through user-defined parameters.

Proper cell spacing near the wall is essential for calculating accurate heating rates. The wall cell spacing in LAURA is controlled by specifying the wall cell Reynolds number:

$$Re_w = \left(\frac{\rho a \Delta \eta}{\mu}\right) \qquad (1)$$

where a is the local sound speed, $\Delta\eta$ is the cell height at the wall and μ is the local viscosity. Grid sensitivity analysis showed that, as seen in Fig. 5, $Re_w = 10$ is sufficient to resolve the laminar boundary heating rate. Requirements for DPLR are similar. The LAURA solutions presented in this work was obtained using $Re_w = 1$.

Fig. 6 shows comparison of convective heating rates along the surface using different diffusion model and chemical species. 12 species mixture is simply the 18 species mixture mentioned above without the ions and electron. The figure shows a variation of approximately 25% in stagnation point heating rate depending on the diffusion model and the chemical species. The two LAURA solutions with 18 species show that a higher stagnation heating rate of 50 W/cm^2 is predicted with the multi-component diffusion model, compared to the stagnation heating rate of 42 W/cm^2 computed with constant Schmidt number. Note the good agreement in the heating rate distribution between the two codes when the same diffusion model of constant Schmidt number of 0.5 and the total number of species are used.

Fig. 7 shows the variation of the translational temperature profile along the stagnation streamline calculated for the nominal case by DPLR and LAURA with three different chemical kinetic models. The forward reaction rate coefficients for each model are listed in Table 2. The backward reaction rates were calculated with equilibrium constants for both codes. Although all the solutions shown predicted the same stagnation point heating rate of approximately 50 W/cm^2, the plot shows that the shock stand off distance and the peak post shock temperature vary with different chemical kinetic models. The profile labeled LAURA (Nelson '91) was computed with LAURA, assuming heavy particles as the collision partner for the ionization reactions, as listed by Nelson.[3] During the study, it was confirmed through a private conversation with the author that the ionization reactions in the article were not appropriate, and that the collision partner should be electrons and not heavy particles. The profile labeled LAURA (Park '85) was computed with the same reaction rates as Nelson '91 but with the correct electron ionization reactions. Comparison of the two profiles shows that the heavy particle ionization reactions contribute to a decrease in both the post shock temperature and shock stand off distance, which is caused by the greater ionization of particles. Comparison of two profiles labeled LAURA (Park '85) and LAURA (Park '89), which are taken from rates published by Chul Park,[13,14] shows the influence of nitrogen dissociation rates on the shock stand off and temperature. The faster nitrogen dissociation rates for Park '89, as shown in Table 2, causes greater dissociation of nitrogen molecules. This decreases the energy within the shock layer that results in slightly lower post shock temperature and shock stand off distance. Lastly, as with the diffusion model comparison, when the same chemical kinetic model Park '85 is used, the stagnation line temperature profiles calculated by the two codes are in relatively good agreement.

Table 3 lists the convective and radiative heating rates at the axisymmetric stagnation point calculated using the three chemical kinetic models. The radiative heating rates were calculated using LORAN[15] with the flow field results from the LAURA calculations. The existence of Cyano radical, CN, contributes to the large values of radiative heating rates, which are approximately twice the amount of convective heating rates. The table shows that the convective heating rates are insensitive to the kinetic models, whereas, the radiative heating rates vary by approximately 225%. The results indicate that, as expected, the radiative heating rate is sensitive to the different post shock temperature and shock stand off distance computed by the different models. A difference in shock stand off distance of just 1 cm can result in a radiative heating rate difference of approximately 30%. Similar results were calculated using NEQAIR[16] and DPLR. Details of the radiation calculations are presented in the companion paper.[2]

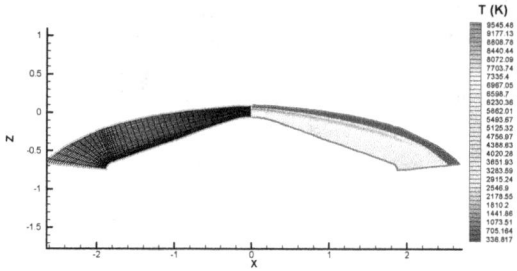

Fig. 4. Temperature contour at peak heating condition.

Table 2. Forward reaction rates.

Reaction	Forward reaction rate coefficient k_f [cm^3/mole-s]
Dissociation Reactions $T_a = (T\, T_v)^{0.5}$	

Nelson '91, Park '89, Park '85

Reaction	k_f
$C_2 + M \rightarrow C + C + M$	$9.68 \times 10^{22}\, T_a^{-2.0} \exp(-71{,}000/T_a)$
$CH + M \rightarrow C + H + M$	$1.13 \times 10^{19}\, T_a^{-1.0} \exp(-40{,}193/T_a)$
$CN + M \rightarrow C + N + M$	$1.00 \times 10^{23}\, T_a^{-2.0} \exp(-90{,}000/T_a)$
$CH_4 + M \rightarrow CH_3 + H + M$	$2.25 \times 10^{27}\, T_a^{-1.87} \exp(-52{,}900/T_a)$
$CH_3 + M \rightarrow CH_2 + H + M$	$2.25 \times 10^{27}\, T_a^{-1.87} \exp(-54{,}470/T_a)$
$CH_2 + M \rightarrow CH + H + M$	$2.25 \times 10^{27}\, T_a^{-1.87} \exp(-50{,}590/T_a)$
$NH + M \rightarrow N + H + M$	$1.13 \times 10^{19}\, T_a^{-1.0} \exp(-41{,}820/T_a)$
$H_2 + M \rightarrow H + H + M$	$1.47 \times 10^{19}\, T_a^{-1.23} \exp(-51{,}950/T_a)$

Nelson '91, Park '85

Reaction	k_f
$N_2 + M \rightarrow N + N + M$	$3.70 \times 10^{21}\, T_a^{-1.6} \exp(-113{,}200/T_a)$

Park '89

$N_2 + N_2 \rightarrow N + N + N_2$	$7.0 \times 10^{21}\, T_a^{-1.6} \exp(-113{,}200/T_a)$
$N_2 + N \rightarrow N + N + N$	$3.0 \times 10^{22}\, T_a^{-1.6} \exp(-113{,}200/T_a)$

Exchange Reactions ($T_a = T$)

Nelson '91, Park '89, Park '85

Reaction	k_f
$C + N_2 \rightarrow CN + N$	$1.11 \times 10^{14}\, T_a^{-0.11} \exp(-23{,}000/T_a)$
$CN + C \rightarrow C_2 + N$	$3.00 \times 10^{14}\, T_a^{0.0} \exp(-18{,}120/T_a)$
$C_2 + N_2 \rightarrow CN + CN$	$7.10 \times 10^{13}\, T_a^{0.0} \exp(-5{,}330/T_a)$
$H + N_2 \rightarrow NH + N$	$2.20 \times 10^{14}\, T_a^{0.0} \exp(-71{,}370/T_a)$
$H_2 + C \rightarrow CH + H$	$1.80 \times 10^{14}\, T_a^{0.0} \exp(-11{,}490/T_a)$
$CN^+ + N \rightarrow CN + N^+$	$9.80 \times 10^{12}\, T_a^{0.0} \exp(-40{,}700/T_a)$
$C + N \rightarrow CN^+ + e^-$	$1.00 \times 10^{15}\, T_a^{1.5} \exp(-164{,}400/T_a)$
$C^+ + N_2 \rightarrow N_2^+ + C$	$1.11 \times 10^{14}\, T_a^{-0.11} \exp(-50{,}000/T_a)$

Associative Ionization Reactions ($T_a = T$)

Reaction	k_f
$N + N \rightarrow N_2^+ + e^-$	$1.79 \times 10^{09}\, T_a^{0.77} \exp(-67{,}500/T_a)$

Ionization Reactions ($T_a = T_v$)

Nelson '91

$N + M \rightarrow N^+ + e^- + M$	$2.50 \times 10^{34}\, T_a^{-3.82} \exp(-168{,}600/T_a)$
$C + M \rightarrow C^+ + e^- + M$	$3.90 \times 10^{33}\, T_a^{-3.78} \exp(-130{,}000/T_a)$
$H + M \rightarrow H^+ + e^- + M$	$5.90 \times 10^{37}\, T_a^{-4.0} \exp(-157{,}800/T_a)$
$Ar + M \rightarrow Ar^+ + e^- + M$	$2.50 \times 10^{34}\, T_a^{-3.82} \exp(-181{,}700/T_a)$

Park '85, Park '89

$N + e^- \rightarrow N^+ + e^- + e^-$	$2.50 \times 10^{34}\, T_a^{-3.82} \exp(-168{,}600/T_a)$
$C + e^- \rightarrow C^+ + e^- + e^-$	$3.90 \times 10^{33}\, T_a^{-3.78} \exp(-130{,}000/T_a)$
$H + e^- \rightarrow H^+ + e^- + e^-$	$5.90 \times 10^{37}\, T_a^{-4.0} \exp(-157{,}800/T_a)$
$Ar + e^- \rightarrow Ar^+ + e^- + e^-$	$2.50 \times 10^{34}\, T_a^{-3.82} \exp(-181{,}700/T_a)$

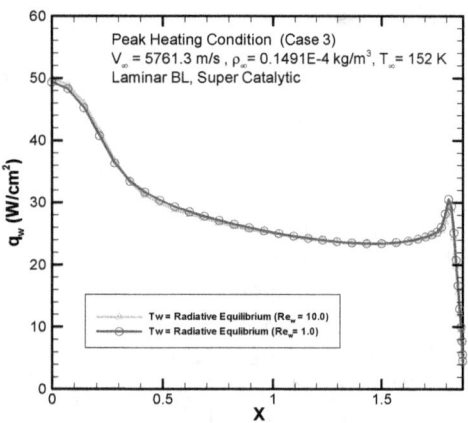

Fig. 5. Wall cell Reynolds number sensitivity.

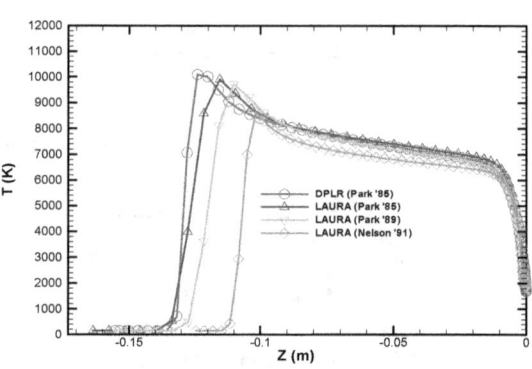

Fig. 7. Translational temperature profile along the stagnation streamline.

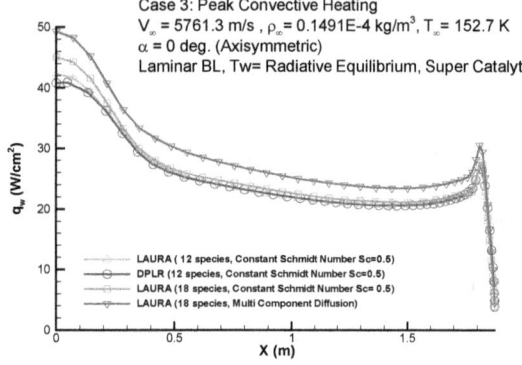

Fig. 6. Variation of convective heating rate along the forebody surface.

Table 3. Axisymmetric stagnation point heating rates for peak heating condition.

	Convective (W/cm2)	Radiative (W/cm2)
		Figure 1. **60.93**
LAURA (Nelson '91)	49.32	
LAURA (Park '89)	51	107.81
LAURA (Park '85)	51.86	138.52

THREE –DIMENSIONAL RESULTS

Three-dimensional calculations were performed along the minimum atmosphere lift-up trajectory for the entry speed of 6.5 km/s using LAURA with 18 chemical species. Multi-component diffusion model with Park '89 chemical kinetic rates were used for the calculation. Fig. 8 shows the multi-block structured grid used for the LAURA calculations.

Fig. 9 shows the shock structure in the pitch plane for the freestream conditions along the 6.5 km/s reference trajectory listed in Table 1. All computations were made for an angle of attack of 16 deg., which is the design trim angle for the aeroshell. Fig. 10 shows the variation of convective heating rate along the surface in the pitch plane for all seven trajectory points, and **Fig. 11** shows the variation of convective heating rate at the nose along the trajectory. The maximum heating rate along the trajectory is 46 W/cm^2, which transpire at the Case 3 peak heating condition. For all cases, the maximum heating rate occurs not at the flow stagnation point, which is located approximately at X= -.98 m, but at the windward side shoulder and the nose/geometric stagnation point. Fig. 12 shows the variation of Re_θ along the forebody pitch plane for all seven cases. Based on a conservative engineering transition criterion of Re_θ of 200, the flow will likely transition to turbulent on the leeward side of the forebody some time before t=253 s (Case 3) and remain turbulent until after t=339 s (Case 6).

Fig. 13 shows variation of convective heating rate along the forebody pitch plane for fully laminar and turbulent flows for Cases 3 and 5. Turbulent flow was modeled using the algebraic turbulence model of Baldwin and Lomax.[17] The figure shows that the turbulent heating rate is 2.5 times that of the laminar heating rate on the leeward side, and the location of maximum heating rate has shifted from the nose and the windward side shoulder to the leeward side shoulder. Although the engineering criterion predicts the flow will be laminar on the windward side, a 10% increase in heating level is observed due to turbulent flow for both cases.

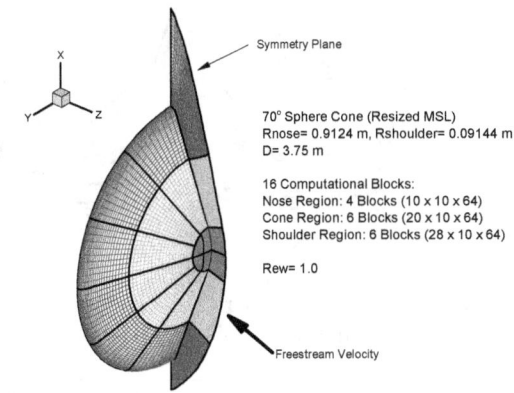

Fig. 8. Computational grid for the forebody calculation

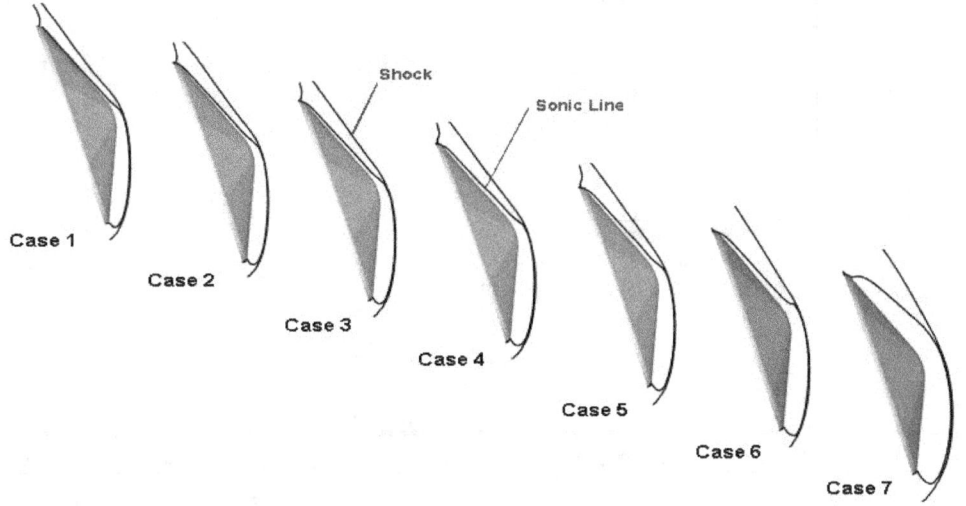

Fig. 9. Shock structure in the pitch plane.

Table 4 lists the lift and drag coefficient, as well as the lift-to-drag ratio of the forebody at each trajectory points. Case 1 data is excluded from the table based on the finding that condition is above the limits for continuum Navier-Stokes calculation. The table shows that the configuration meets the preliminary L/D design requirement of 0.25 along the trajectory. The CG location at the 16° trim angle of attack is at z/D = 0.3 and x/D = 0.0197.

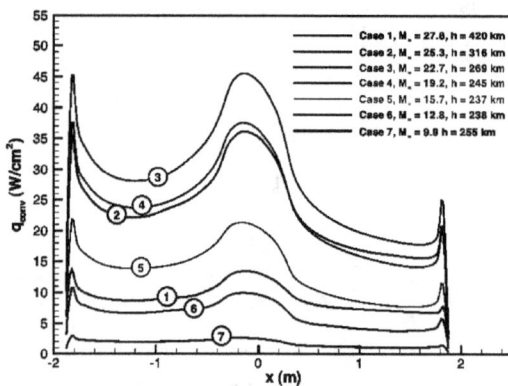

Fig. 10. Variation of convective heating rate on the forebody for the seven trajectory points.

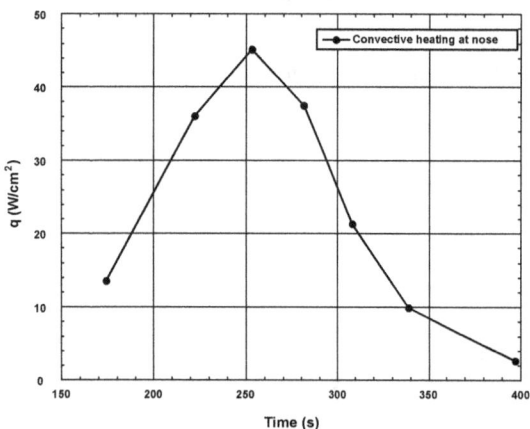

Fig. 11. Variation of convective heating rate along the reference trajectory.

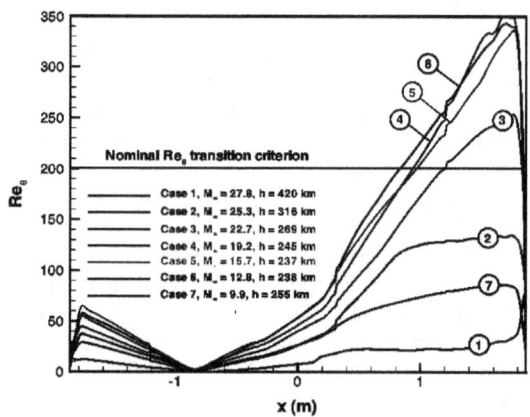

Fig. 12. Variation of Re_θ along the forebody for each trajectory point.

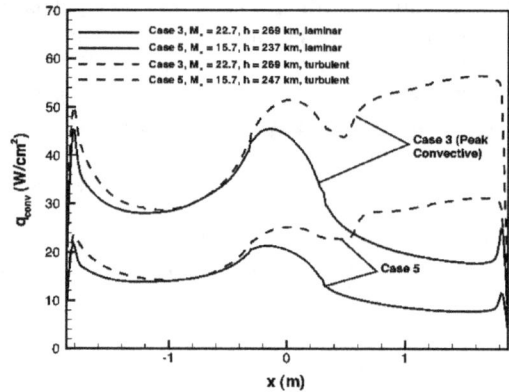

Fig. 13. Turbulent and laminar convective heating rates for Case 3 and 5.

Table 4. Vehicle aerodynamics.

	Case 1	Case 2	Case 3	Case 4	Case 5	Case 6	Case 7
CL	*	0.360	0.360	0.357	0.357	0.360	0.362
CD	*	1.446	1.437	1.421	1.420	1.427	1.429
L/D	*	0.249	0.251	0.252	0.251	0.252	0.254

SUMMARY AND CONCLUSIONS

Aeroheating environments along a reference trajectory were calculated for the Titan Aerocapture System Analysis study. The environment was calculated using two different CFD codes by two teams from different NASA Research centers for the purpose of reducing uncertainties through direct comparison of results. Sensitivity analyses showed that two codes are in good agreement when identical transport/chemical kinetic models are used. Axisymmetric flow calculations showed that while insensitive to the chemical kinetic rates, convective heating rates are sensitive to the diffusion model and the choice of chemical species. In contrast, the radiative heating rates calculated using the CFD solutions are sensitive to the shock stand off distance and the post shock temperature profile, which are influenced by the chemical kinetic model as expected. Radiation calculations using the CFD results showed that the majority of the heating on the aeroshell will be due to radiative heating, which will drive not only the TPS sizing but also the material selection due to the nature of radiative heating. For this reason, careful reexaminations of the chemistry models are needed to reduce the uncertainties in the radiative heating.

Three dimensional forebody calculations showed that the reference configuration meets the design L/D requirement of 0.25 along the reference trajectory. Momentum thickness Reynolds number along the forebody indicates that the flow is likely to transition to turbulent on the leeward side. Turbulent calculations show that the convective heating level on the leeside is increased by 250%, and the location of the maximum heating moves from the nose and the windward shoulder to the leeside shoulder.

ACKNOWLEDGEMENTS

The authors would like to acknowledge Peter Gnoffo for his valuable contribution towards the sensitivity analyses and assistance with the LAURA CFD program. In addition, the authors would like to acknowledge Lin Chambers for her assistance with the LORAN radiation program.

Naruhisa Takashima is supported under NASA Langley Research Center SAMS contract NAS1-00135. Ken Sutton is supported under contract NAS1-02117 from NASA Langley Research Center to National Institute of Aerospace. Michael Wright and Dinesh Prabhu were supported under contract NAS2-99092 from NASA Ames Research Center to ELORET Corp.

REFERENCES

[1] Lockwood, M, "Titan Aerocapture Systems Analysis," AIAA 2003-4799, July 2003.

[2] Olejniczak, J., et al., "An Analysis of the Radiative Heating Environment for Aerocapture at Titan," AIAA 2003-4953, July 2003.

[3] Nelson, H. F., Park, C. and Whiting, E. E., "Titan Atmospheric Composition by Hypervelocity Shock-Layer Analysis," *Journal of Thermophysics*, Vol. 5, No. 2, April-June 1991, pp 157-165.

[4] Baillion, M., Pallegoix, J. F. and Soler, J., "Huygens Probe Aerothermodynamics," AIAA 97-2476, June 1997.

[5] Park, C. and Bershader D., "Determination of the Radiative Emission of a Hypersonic Flow Simulating the Cassini-Titan Atmospheric Entry Probe Environment," AIAA 90-1558, June 1990.

[6] Baillion, M., "Aerothermodynamic Requirements and Design of the Huygens Probe," *Capsule Aerothermodynamics*, AGARD-R-808, May 1997.

[7] Edquist, K. T., et al., "Aeroheating Environments for a Mars Smart Lander," AIAA 2002-4505, Aug. 2002.

[8] Justus, C. and Duvall, A., "Engineering- Level Model Atmospheres for Titan and Neptune," AIAA-2003-4803, July 2003.

[9] Way, D., Powell, R. and Edquist, K., "Aerocapture Simulation and Performance for the Titan Explorer Mission," AIAA 2003 4951, July 2003.

[10] Gnoffo, P. A., "Computational Aerothermodynamics in Aeroassist Applications," AIAA 2001-2632, 2001.

[11] Wright, M. J., Candler, G. v. and Bose, D., "Data-Parallel Line Relaxation Method for Navier-Stokes Equations," *AIAA Journal*, Vol. 36, No. 9, Sept. 1998.

[12] Wright, M.J., Loomis, M., and Papadopoulos, P., 'Aerothermal Analysis of the Project Fire II Afterbody Flow,' *Journal of Thermophysics and Heat Transfer*, Vol. 17, No. 2, pp. 240-249, 2003.

[13] Park, C. "On Convergence of Chemically Reacting Flows" AIAA 85-0247, Jan. 1985.

[14] Park, C. "A Review of Reaction Rates in High Temperature Air" AIAA 89-1740, June 1989.

[15] Chambers, L. H., "Predicting Radiative Heat Transfer in Thermochemical Nonequilibrium Flow Fields," NASA TM-4564, Sept. 1994.

[16] Whiting, E. E., et al., "NEQAIR96, Nonequilibrium and Equilibrium Radiative Transport and Spectra Program: User's Manual," NASA RP-1389, Dec. 1996.

[17] Cheatwood, F. M. and Thompson, R., A., "The Addition of Algebraic Turbulence Modeling to Program Laura," NASA TM 107758, April 1993.

AN ANALYSIS OF THE RADIATIVE HEATING ENVIRONMENT FOR AEROCAPTURE AT TITAN

Joseph Olejniczak
NASA Ames Research Center, Moffett Field, CA 94035
Michael Wright, Dinesh Prabhu
ELORET Corp., 690 W. Fremont Ave., Suite 8, Sunnyvale, CA 94087
Naruhisa Takashima
AMA Inc., 303 Butler Farm Road, Suite 104A, Hampton, VA 23666
Brian Hollis, E. Vincent Zoby,
NASA Langley Research Center, Hampton, VA 23681
and
Kenneth Sutton
National Institute of Aerospace, Hampton, VA 23666

ABSTRACT

Details of the radiative heating analysis for the forebody of a candidate Titan aerocapture orbiter are presented. The radiative heating rates are obtained through *a posteriori* analysis of high-fidelity thermochemical nonequilibrium flow fields computed using modern techniques of computational fluid dynamics. Results from axisymmetric and three-dimensional analysis are presented at several points on candidate aerocapture trajectories in various model atmospheres of Titan. The radiative heating rates are found to be up to five times the peak convective heating rates, indicating that an accurate knowledge of the uncertainty of the radiative heating predictions is critical for a Titan aerocapture mission. The results also show that – (1) the radiative heating rates are dominated by the violet band system of CN, and (2) the gas mixture is optically thin. The predicted radiative heating is found to be very sensitive to the dissociation rate of molecular nitrogen – a factor of two increase in the rate, which is within the experimental uncertainty, results in a 25% decrease in the radiative heating.

INTRODUCTION

A one-year system analysis design study[1,2] of an aerocapture mission to the Saturn moon Titan was recently completed. The purpose of the systems study was to identify the critical issues for an aerocapture mission, not necessarily to provide a complete mission design. The purpose of the radiation calculations presented in this paper was likewise to assess the impact of radiative heating on the system design and identify the further work needed before a Titan aerocapture mission could be launched.

Aerocapture uses a single pass through an atmosphere to dissipate excess energy and enter a stable orbit.

Aerocapture missions offer a mass savings over traditional propulsive capture missions if the mass fraction of the Thermal Protection System (TPS) needed to protect the vehicle during the atmospheric pass is less than the mass fraction of the propulsion system needed to slow the vehicle and enter orbit. The results of the systems analysis study[1] demonstrated that aerocapture missions at Titan provide a significant mass benefit compared to propulsive capture missions.

A 70-degree sphere-cone with a 1.875 m maximum radius was selected as the baseline design for the orbiter, as shown in Fig. 1. The geometry is a scaled version of the proposed Mars Smart Lander vehicle.[3] For the nominal entry trajectory, the vehicle enters the atmosphere with a relative velocity of 6.5 km/s and flies at a constant 16° angle-of-attack. For comparable velocities and altitudes at Earth, the corresponding radiative heating level would be insignificant.

However, studies of the heating environment of the Titan probe Huygens indicated that radiative heating rates were 2-3 times greater than convective heating rates.[4-6] The Huygens probe, part of the Cassini mission to Saturn, is a 2.70 (m) diameter 60-degree sphere cone that will enter the Titan atmosphere on a direct, ballistic trajectory. Since the proposed aerocapture orbiter is larger than Huygens and flies a longer, lifting trajectory, it was expected that the aerocapture orbiter would experience even higher radiative heating rates than the Huygens probe.

Titan has an atmosphere of nitrogen, methane, and argon. This gas composition leads to non-intuitive thermochemical nonequilibrium effects in the shock layer. The methane in the atmosphere dissociates in the nonequilibrium shock layer and CN forms. The CN molecule is a strong radiator. Therefore, even though the

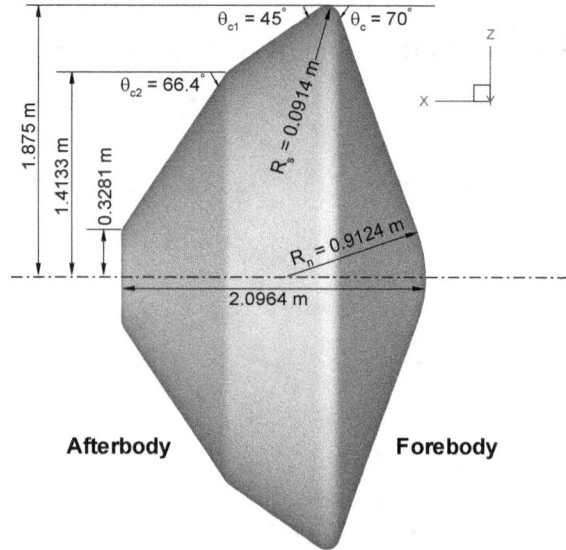

Figure 1. Schematic view of the Titan aerocapture orbiter.

entry velocity and post-shock temperatures are relatively low compared to Earth entry, there is the potential for significant nonequilibrium radiation.

The gas kinetic and radiation models used in this work are based on the work of Nelson et al.,[7] which was developed using shock tube data from Park.[8] The Park data are from only one velocity-pressure condition. A detailed collisional-radiative model based on extensive shock tube data was developed to design the Huygens probe,[4-6] but details of the experimental data and the model have not been published.

Because the radiative heating has a first order impact on the TPS material selection and mass, even a preliminary design of an aerocapture vehicle requires reliable estimates of the radiative heating. As a result, high fidelity CFD and radiation calculations using the best available thermophysical models were made to support the system design study.

This paper discusses these radiative heating predictions for the forebody of the Titan aerocapture orbiter. First, a description of the methods used to compute the radiative heating rates is given. Second, radiative heating results for axisymmetric and three-dimensional cases along candidate aerocapture trajectories are presented. Finally, the status of the ongoing efforts is reported. A series of companion papers describes the details and results of the mission design.[9-16] Takashima et al.[17] discusses the convective aeroheating environment for the aerocapture vehicle.

METHODOLOGY

The flow field and radiation production and transport were calculated in an uncoupled manner. In other words, Computational Fluid Dynamics (CFD) was used to compute the flow field properties used as input to the radiation solvers, but the effect of the radiation on the flow field was neglected. Previous authors[18,19] have used loosely or fully-coupled methods for strongly radiating flows. However, the uncoupled approach is appropriate for systems analysis trade studies because it provides a conservative estimate of the radiative heating at a much lower computational cost than a loosely or fully coupled approach, allowing a large number of candidate designs to be rapidly analyzed. The degree to which the uncoupled approach over-predicts the total radiative heating for this application will be discussed in a later section.

The radiation calculations are made by computing the radiative flux that reaches a location on the surface of the vehicle along a line-of-sight normal to that surface location. The calculations required as inputs the species number densities, translational temperature, and vibrational temperature at each point in the flow field along the line-of-sight. The radiation solvers compute the amount of radiation produced and the transport of that radiation along each line-of-sight.

CFD

Two different CFD solvers were used to compute the flow fields: DPLR[19] and LAURA.[20] Briefly, both codes solve the Navier-Stokes equations including finite rate chemistry and vibrational nonequilibrium. DPLR and LAURA have been shown to give essentially the same results when using identical thermophysical models.[17]

The chemical reaction rates used were from Nelson et al.[7] The rate model in Nelson et al. was developed for application to Titan entry and was based on a combination of shock tube data from Park[8] and existing reaction rate models. Because the amount of nitrogen dissociation has a direct impact on the amount of CN formed, some calculations were made substituting the newer nitrogen dissociation rates from Park[22] in order to test the sensitivity of the radiative heating to the reaction rate model. The rates in Park[22] are approximately twice as fast as those in the Nelson et al. model. The effect of reaction rate choice on radiative heating will be discussed in detail in a later section. Most calculations were made using a twelve species model (CH_4, CH_3, CH_2, N_2, C_2, H_2, CH, NH, CN, N, C, and H). Calculations were also made using an eighteen species (CH_4, CH_3, CH_2, N_2, N_2^+, C_2, H_2, CH, NH, CN, CN^+, N, N^+, C, C^+, H, H^+, and e) to confirm that ionization effects were unimportant.

RADIATION

Similarly, two radiation transport codes were used: NEQAIR96[23] and LORAN.[24] Both of these radiation solvers calculate the emission and transport of radiation in the flow field using the tangent slab approximation. In this approximation radiation is assumed to be emitted from infinitely thin parallel planes perpendicular to the absorbing surface. NEQAIR96 performs the calculations line-by-line, computing an absorption coefficient for each atomic line

Table 1. Methane mass fractions for the three different Titan atmosphere models.

Atmosphere	CH$_4$ Mass Fraction
Minimum	0.0293
Nominal	0.0173
Maximum	0.0055

and molecular rotational line in the vibrational bands. LORAN uses a smeared band model with a smoothly varying absorption coefficient for the molecular band. Comparisons of NEQAIR96 and LORAN results will be presented later.

The tangent slab results computed in either NEQAIR96 or LORAN need to be reduced to account for surface curvature effects. The exact value of this reduction factor depends on the surface geometry and the absorption in the flow field. For a spherical stagnation point in air, values from 0.75 to 0.85 have been previously used.[25,26] In this work, a value of 0.80 was chosen.

The specific molecular bands considered were N$_2$ 1+ (B $^3\Pi_g$->A $^3\Sigma_u^+$), N$_2$ 2+ (C $^3\Pi_u$->B $^3\Pi_g$), N$_2$ Lyman-Birge-Hopfield (a $^1\Pi_g$->X $^1\Sigma_g^+$), CN violet (B $^2\Sigma^+$->X $^2\Sigma^+$), CN red (A $^2\Pi$->X $^2\Sigma^+$), H$_2$ Werner (C $^1\Pi_u$->X $^1\Sigma_g^+$), and H$_2$ Lyman (B $^1\Sigma_u^+$->X $^1\Sigma_g^+$). Also computed was radiation from atomic lines of N, C, and H. Updated molecular spectroscopic data from Laux[27] and atomic spectroscopic data from NIST[28] were used in NEQAIR96.

While CFD calculates the number of molecules and atoms in the ground electronic state, radiation is emitted from excited electronic states of molecules and atoms. Therefore, a collisional-radiative excitation model is needed to compute the number of electronically excited molecules and atoms and their distribution among vibrational and rotational states.

Since no validated excitation model was available for a nitrogen, methane, and argon gas mixture, the excited electronic states of the molecules and atoms were assumed to be in equilibrium, and therefore populated in a Boltzmann distribution governed by the vibrational temperature of the gas. While such an assumption is inappropriate for air mixtures, it is a reasonable approximation when CN is the main radiator emitting in the violet and red bands. Since the ground (X $^2\Sigma^+$), first excited (A $^2\Pi$), and second excited (B $^2\Sigma^+$) states of the CN molecule lie close together and are distinct with no crossings, it is a reasonable first approximation to assume that gas collisional rates will be sufficiently high to maintain a Boltzmann distribution among the first three electronic states.

TRAJECTORIES

The systems analysis group provided four different design trajectories, V6.5.[2] The aerothermal environments were computed at seven to nine specific points along each trajectory. Freestream conditions were taken from an atmospheric model[8] developed as part of the Titan mission study. Both axisymmetric and three-dimensional CFD

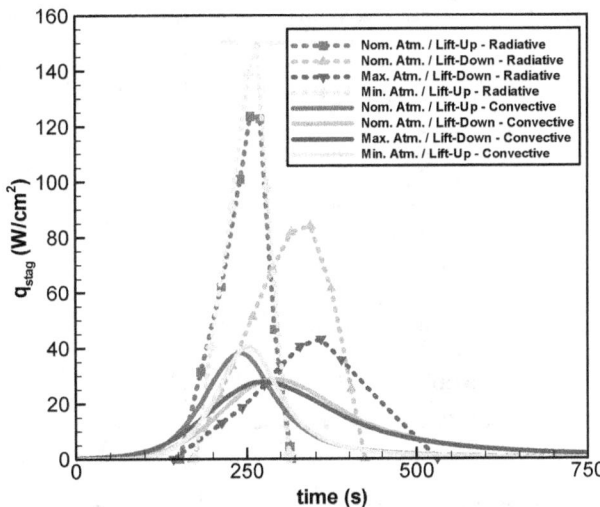

Figure 2. Predicted axisymmetric convective and radiative heating rates for four entry trajectories.

calculations were made. Radiative heating values were computed at selected points on the body including the stagnation point.

RESULTS

For aerocapture missions to Titan with entry velocities below about 8.5 km/s, our results show that over 90% of the radiative heating is from the CN[v] band. The remaining radiation is from the CN red band (CN[r]), various N$_2$ molecular systems, and N atomic lines. Because the radiation from CN[v] is relatively short-wavelength, 340–430 nm, there is a concern that it will penetrate a low density TPS material and heat the material in-depth.[8] The NEQAIR96 radiation calculations also showed that over the entire range of conditions of interest for Titan aerocapture the gas is optically thin; i.e. there is no absorption of the radiation by the gas in the flow field.

AXISYMMETRIC CASES

Even though the aerocapture vehicle flow field is three-dimensional, initial calculations were made on the orbiter geometry assuming an axisymmetric flow field. In this way, a large number of trajectory points can be analyzed to provide radiative heating sensitivities to the design variables being studied by the systems analysts. Figure 2 shows the computed axisymmetric stagnation point convective and radiative heating rates for four entry trajectories corresponding to three different atmosphere models and two different flight path angles.

Since the exact composition of the Titan atmosphere is unknown, three different atmospheric models are considered: minimum, nominal, and maximum, where the nomenclature refers to the atmospheric density at a given altitude.[10] It is further assumed that the different atmospheric models predict differing amounts of methane.

Table 2. Predicted axisymmetric peak radiative heating rates and radiative heat loads for four entry trajectories.

Trajectory	Peak Radiative Heating (W/cm^2)	Radiative Heat Load (J/cm^2)	Peak Convective Heating (W/cm^2)	Convective Heat Load (J/cm^2)
Nominal Atm. *Lift-Up*	124	10200	39	5500
Nominal Atm. *Lift-Down*	84	12090	29	7500
Maximum Atm. *Lift-Down*	43	8400	28	7700
Minimum Atm. *Lift-Up*	149	10580	41	5200
Minimum Atm. *Lift-Down*	100	12480	30	7600

The atmosphere-specific methane mass fractions are listed in Table 1. Lift-up indicates a steep entry angle requiring full lift to capture to the desired orbit. Lift-down indicates a shallow entry angle requiring full lift-down to capture to the desired orbit. The lift-up and lift-down trajectories thus define the boundaries of the entry corridor. The lift of the vehicle is controlled by roll modulation while at constant angle-of-attack.

The corresponding stagnation point heat loads and values of peak convective and radiative heating rates are given in Table 2. The convective heating rates are below 50 W/cm^2, however the radiative heating rates are up to 150 W/cm^2 for the minimum atmosphere, lift-up trajectory. Peak radiative heating occurs after peak convective heating. For the minimum atmosphere, lift-up trajectory peak convective heating is at $t=253$ s and peak radiative heating is at approximately $t=266$ s. The largest radiative heat load occurs on the minimum atmosphere, lift-down trajectory, and the largest total heat load is for the nominal atmosphere, lift-down trajectory.

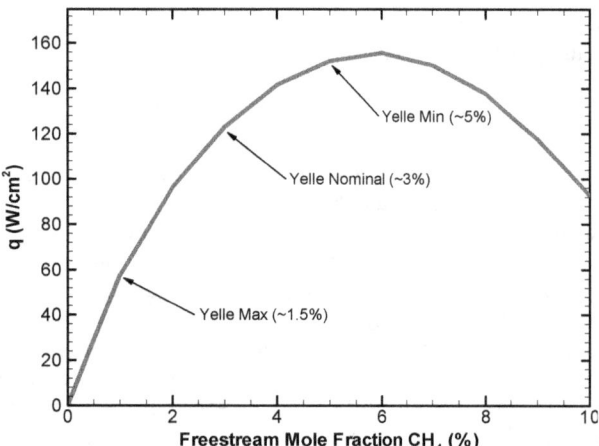

Figure 3. Effect of assumed freestream methane mole fraction on axisymmetric stagnation point radiative heating predictions for the $t = 253$ s point of the $v = 6.5$ km/s entry minimum atmosphere, lift-up trajectory.

Interestingly, the trajectory with the peak convective heating rate is the trajectory with the peak radiative heating rate, yet the trajectory with the maximum convective heat load is not the trajectory with the maximum radiative heat load. This surprising result is because the convective heating depends on the freestream enthalpy, while the radiative heating depends on temperature and CN concentration distributions through the entire shock-layer.

Although not shown, the predicted radiative heating rates are higher along the conical flank of the orbiter than at the stagnation point. The peak value is about 230 W/cm^2. The post-shock temperatures and CN mass fractions are similar in the two regions, and the heating is higher in the flank region simply because there is a larger volume of radiating gas.

The results shown here are from the DPLR CFD code and the NEQAIR96 radiation code using the Nelson *et al.* rate model. The results using the LAURA CFD code and the LORAN radiation code with the Nelson *et al.* rate model are within 5% of the DPLR/NEQAIR96 results. The radiative heating rates are from CN[v] only. The impact of including CN[r], molecular nitrogen bands, and atomic nitrogen lines is discussed below.

Because of chemical nonequilibrium in the shock layer, the amount of CN radiation depends strongly on the freestream methane concentration. At equilibrium, the shock-layer temperatures would be too high for CN to exist since the dissociation energy is only 7.65 eV. Figure 3 shows the calculated radiative heating rates as a function of assumed freestream methane concentration for the $t=253$ s point on the minimum atmosphere, lift-up trajectory. The freestream velocity is 5761 m/s and the density is 1.491e-4 kg/m^3. The amounts of methane corresponding to the three atmosphere models discussed above are indicated on the figure. The amount of methane predicted by the minimum atmosphere model happens to result in nearly the peak amount of radiative heating, indicating that designing the vehicle using the minimum atmosphere is the conservative choice for radiative heating.

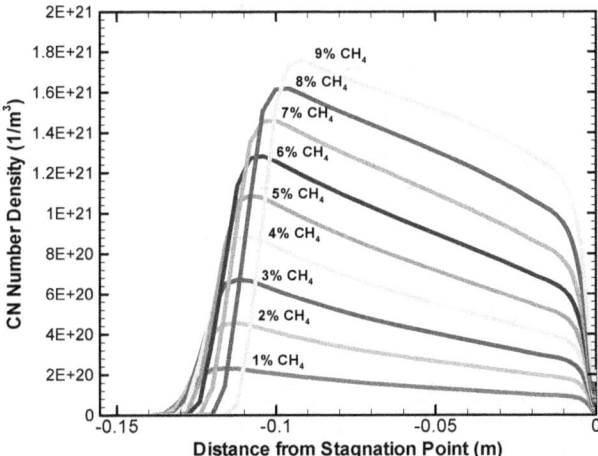

Figure 4. Effect of assumed freestream methane mole fraction on post-shock CN number density.

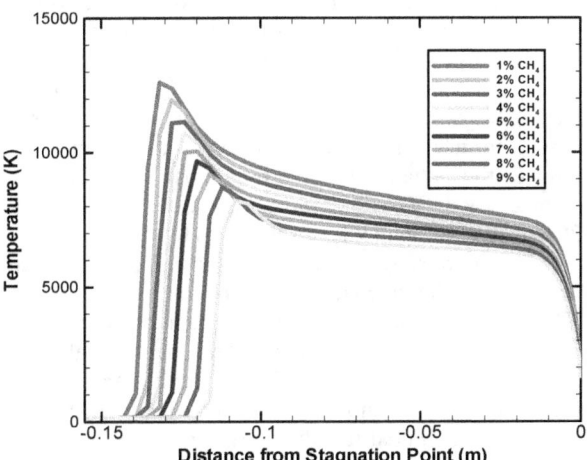

Figure 5. Effect of assumed freestream methane mole fraction on post-shock translational temperature.

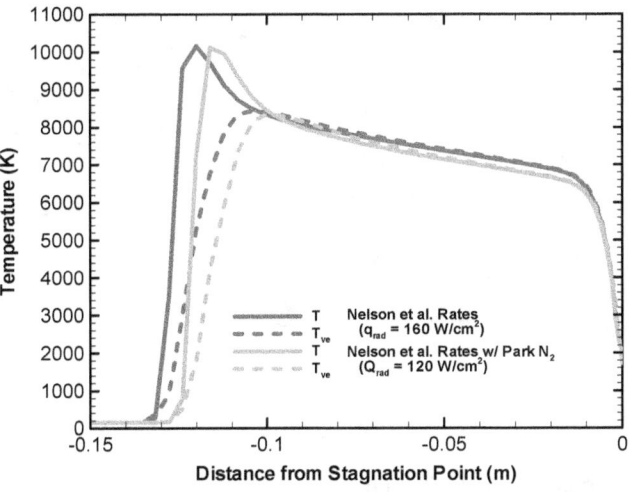

Figure 6. Computed axisymmetric stagnation line temperature profiles for two different reaction rate models.

The methane dissociates into its constituent atoms immediately downstream of the shock and quickly forms CN through various exchange reactions with the atmospheric nitrogen. More methane available means more CN is produced, as shown in Fig. 4. Therefore, at low methane concentrations the amount of radiation is proportional to the amount of methane in the atmosphere. As the amount of methane increases another effect becomes important, namely that the increased amount of methane dissociation decreases the post-shock temperature (Fig. 5). Even though there are more CN molecules, the effective temperature at which the CN radiates is lower. The net effect is to decrease the amount of CN radiation since the radiation intensity depends linearly on the number of radiating molecules but exponentially on the vibrational temperature. The vibrational temperature is in equilibrium with the translational temperature for about 70% of the shock layer.

There is some uncertainty in the thermophysical models for the Titan atmosphere. One important rate is the nitrogen dissociation rate. Figure 6 shows the effect of the value of the nitrogen dissociation rate on the shock stand-off distance and the resulting effect on the computed radiative heating rate. The two rates differ only by a constant factor of about two, and yet the predicted radiative heating rates differ by 25%. Additional data are needed to reduce this uncertainty.

THREE-DIMENSIONAL CASES

Three-dimensional calculations were also made for the lift-up, minimum atmosphere trajectory, which has the highest radiative heating rates. The distributed radiative heating on the forebody of the vehicle is shown in Fig. 7 for the $t=253$ s point of the lift-up minimum atmosphere trajectory with $\alpha=16°$. Because the gas is optically thin, the radiative heating at the surface is given by the sum of

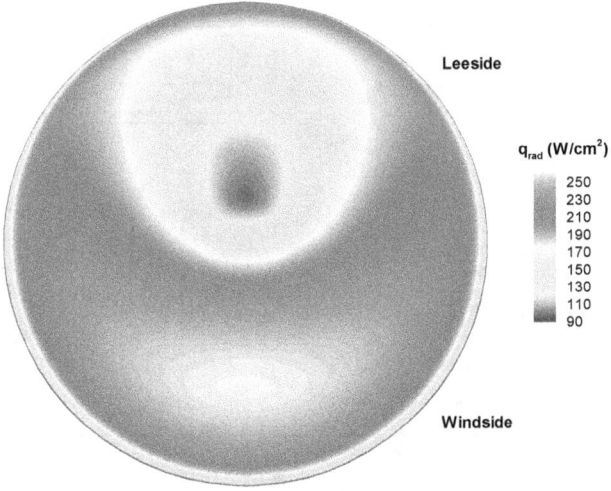

Figure 7. Computed three-dimensional forebody radiative heating rate for the $t = 253$ s point of the $v = 6.5$ km/s entry minimum atmosphere, lift-up trajectory.

Table 3. Three-dimensional stagnation point radiative heating rates for the $t = 253$ s point of the $v = 6.5$ km/s entry minimum atmosphere, lift-up trajectory.

CFD Code	# of Species	Kinetics	q_{rad} (CN Violet) (W/cm^2)	
			NEQAIR96	LORAN
DPLR	12	Nelson et al.	278	246
DPLR	12	Nelson et al. & Park N$_2$	214	
DPLR	18	Nelson et al.	289	
LAURA	18	Nelson et al. & Park N$_2$	202	175

the contributions of the individual radiating volumes of gas along a line-of-sight perpendicular to the surface. The peak radiative heating value is about 260 W/cm^2 is near the flow stagnation point where there is the largest volume of radiating gas. In the axisymmetric case, the largest volume of gas was in the conical flank region and the radiative heating rate is about 230 . W/cm^2. The axisymmetric stagnation point heating rate of about 150 W/cm^2 is similar to the value at the center of the spherical nose for the three-dimensional calculation.

Figure 8 shows a comparison of the convective and radiative heating rates along the pitch-plane of the forebody for the $t=253$ s point of the lift-up minimum atmosphere trajectory. The flow stagnation point is at approximately $y=0.8$ m. The radiative heating is significantly larger than the convective heating over the entire forebody.

Table 3 gives the three-dimensional stagnation point heating rates for the t=253 point of the lift-up minimum atmosphere trajectory for various combinations of CFD code, radiation code, and kinetic model. Using DPLR/NEQAIR96 the difference between the 12 species and 18 species radiative heat transfer is only 4%, indicating that ionization effects can be ignored for the systems studies. Using the DPLR flow field solution and the Nelson et al. model, LORAN predicts approximately 13% less radiation than NEQAIR96. Considering that LORAN uses a lumped band model as opposed to the detailed line-by-line method of NEQAIR96, there is reasonable agreement between the two codes. For this case, the use of the Park nitrogen dissociation rates reduces the radiative heating by almost 30%. The results from DPLR/NEQAIR96 and LAURA/LORAN are within 20% when the same kinetics model is used.

Figure 8 shows that the highest radiative heating occurs near the stagnation point of the vehicle. Since at higher entry velocities the post-shock temperature in the stagnation region will eventually become too high for CN to form, it was thought that increasing the entry velocity might reduce the amount of radiative heating. Figures 9 shows the computed radiative heating on the forebody for an entry velocity of 10 km/s at the peak heating point. The radiative heating levels in the stagnation region are indeed reduced from about 250 W/cm^2 to about 160 W/cm^2.

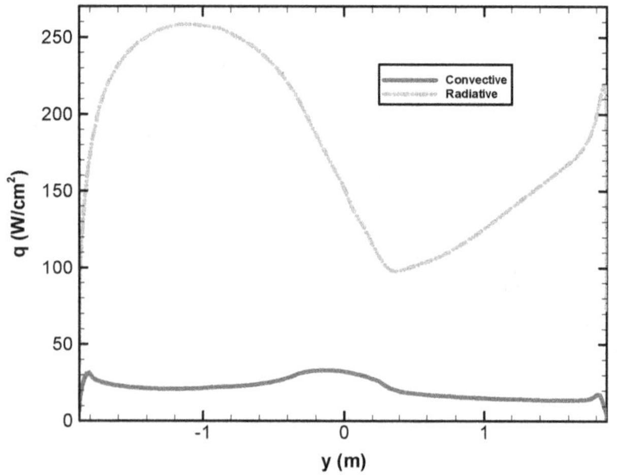

Figure 8. Computed pitch-plane forebody convective and radiative heating rate for the $t = 253$ s point of the minimum atmosphere, lift-up trajectory.

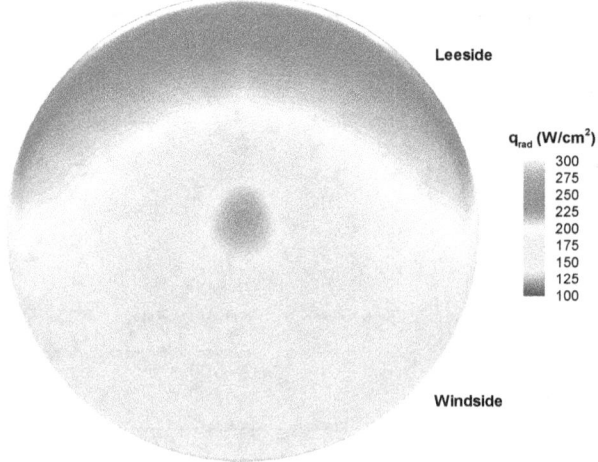

Figure 9. Computed three-dimensional forebody radiative heating rate for the peak convective heating point ($t = 152$ s) of the $v = 10$ km/s minimum atmosphere, lift-up trajectory

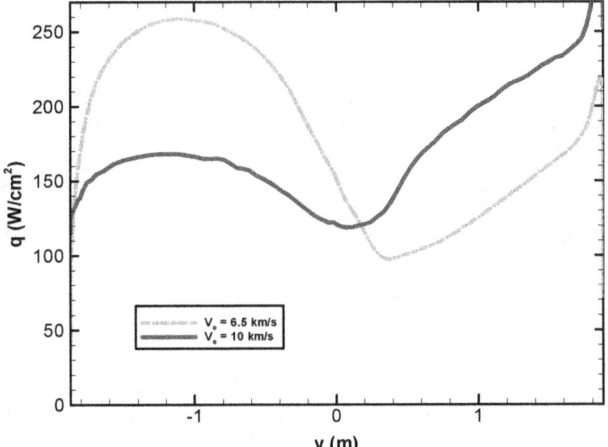

Figure 10. Computed pitch-plane forebody radiative heating rate for the $t = 253$ s point of the minimum atmosphere, lift-up trajectory and the $t = 152$ s point of the $v = 10$ km/s minimum atmosphere, lift-up trajectory.

However, for the 10 km/s entry the leeside of the vehicle has post-shock conditions similar to the 6.5 km/s stagnation point, and the radiative heating rate is as high as 220 W/cm^2. A comparison of the radiative heating rates for the two entry velocities along the pitch-plane is shown in Fig. 10. In fact, the total integrated surface heating rate over the orbiter surface increases by 5%. Furthermore, this analysis does not include the contribution of the other radiators such as N_2 and N, which will contribute to the radiative heating at this entry velocity. Additionally, the convective heating will also be about 3.5 times higher compared to the 6.5 km/s entry trajectory. Therefore, increasing the entry velocity does not reduce the amount of heating to the vehicle.

FUTURE WORK

The prior analyses all assumed that the radiation calculation could be separated (decoupled) from the flow field calculation. This assumption is appropriate if the flow field is adiabatic and the energy lost from radiation is small. However, given the high levels of radiative heating predicted at relatively low entry velocities, it was decided to examine the impact of radiation-fluid coupling on the aerothermal heating predictions.

It is possible to estimate the amount of coupling between the radiation and fluid by evaluating the radiative loss parameter[29], Γ:

$$\Gamma = 2q_{rad}/\tfrac{1}{2}\rho_\infty V_\infty^3 \qquad (1)$$

where q_{rad} is the stagnation point radiative heat flux, ρ_∞ is the freestream density, and V_∞ is the freestream velocity. When Γ becomes large (> 1%), the flow is considered to be non-adiabatic and coupled to the radiation field. If a large amount of flow energy is converted to radiation, the shock-layer is cooled and there is a potentially significant impact on the fluid dynamics and chemical kinetics of the flow. The net effect of the coupling would be to reduce the amount of radiative heating.

The radiative loss parameter for the reference Titan aerocapture vehicle was found to exceed 40% near peak radiative heating, indicating that the flow is non-adiabatic and coupled. By comparison, Γ is about 1% for the Fire-II flight experiment, and about 15% for Galileo. Further work investigating the effect of coupling is ongoing and will be presented in an upcoming paper. Preliminary results show that the coupled radiative heating rates are still 2-3 times greater than the convective heating rates.

Preliminary uncoupled afterbody calculations have also been made in order to assess the amount of radiative heating on the afterbody. The radiative heating analysis is still in progress and will be reported in an upcoming paper.

Finally, because the work presented in this paper identified that radiative heating is a significant design issue for Titan missions, experiments in the NASA Ames EAST shock tube are about to begin. The tests will provide data over a range of velocity and pressure conditions that will be used to either validate the existing chemical kinetic and radiation model or construct a new collisional-radiative model for application to the Titan atmosphere. The development of this new model will allow greater confidence in radiative heating predictions and quantified uncertainty estimates for Titan entry missions.

SUMMARY AND CONCLUDING REMARKS

Axisymmetric and three-dimensional thermochemical nonequilibrium Navier-Stokes calculations using DPLR and LAURA were made for a candidate Titan aerocapture orbiter along four design trajectories. The computed flow fields were used as inputs to the radiation solvers NEQAIR96 and LORAN. The combinations DPLR/NEQAIR96 and LAURA/LORAN gave similar answers.

Radiative heating was found to be a significant design issue for Titan aerocapture missions. The current results are conservative since they do not account for the radiative cooling effect, but this work has identified the key issues and sensitivities for the radiative heating environment.

The radiation results from the nonequilibrium formation of CN in the shock layer around the vehicle created in the nitrogen–methane atmosphere. In fact, if the flow field were in equilibrium, the radiative heating would be negligible compared to the convective heating. The radiation field was found to be optically thin for all conditions relevant to the aerocapture orbiter.

It was found that the peak radiative heating rate is five times greater than the peak convective heating rate, and the radiative heat load is twice as great as the convective heat loads for the minimum atmosphere, lift-up trajectory. Three-dimensional calculations showed that there was significant radiation on the entire forebody.

The predicted radiative heating was found to be sensitive to the choice of the nitrogen dissociation rate. A factor of two change in the rates produced a 25% change in the predicted heating rate. The factor of two difference in the rates is well within the experimental uncertainty of the data, indicating more accurate knowledge of the nitrogen dissociation rate is needed to reduce the uncertainty in the radiative heating predictions.

This work was performed as part of a preliminary system analysis study, and as expected it raised as many questions as it answered. Two of these questions, the effects of strongly coupled radiation and the impact of radiative heating on the afterbody, were mentioned in this paper. This work also identified the need for additional shock tube data in order to develop the collisional-radiative model necessary for a detailed mission design.

ACKNOWLEDGMENTS

M. Wright and D. Prabhu were supported by NASA Ames Research Center, through contract NAS2-99092 to ELORET. N. Takashima was supported by NASA Langley Research Center, through contract to AMA, Inc. K. Sutton was supported by NASA Langley Research Center, through the National Institute of Aerospace, contract number NAS1-02117.

REFERENCES

[1] Lockwood, M.K., "Titan Aerocapture Systems Analysis," AIAA Paper No. 2003-4799, Jul. 2003.

[2] Lockwood, M.K., Sutton, K., Prabhu, R., et al. "Entry Configurations and Performance Comparisons for the Mars Smart Lander," AIAA Paper No. 2002-4407, Aug. 2002.

[3] Bailey, R., Hall, J., and Spilker, T., "Titan Aerocapture Mission and Spacecraft Design Review," AIAA Paper No. 2003-4800, Jul. 2003.

[4] Baillion, M., Pallegoix, J.F., and Soler, J., "Huygens Probe Aerothermodynamics," AIAA Paper No. 97-2476, Jul. 1997.

[5] Baillion, M., Taquin, G., and Soler, J., "Huygens Probe Radiative Environment," *Proceedings of the 19th International Symposium on Shock Waves, Shock Waves at Marseille II: Physico-Chemical Processes and Nonequilibrium Flow*, Springer-Verlag, 1995, p. 339-346.

[6] Baillion, M. and Taquin, G., "Radiative Heat Flux: Theoretical and Experimental Predictions for Titan Entry Probe," *Capsule Aerothermodynamics*, AGARD Report No. 808, May 1997.

[7] Nelson, H.F., Park, C., and Whiting, E.E., "Titan Atmospheric Composition by Hypervelocity Shock-Layer Analysis," *J. Thermophysics & Heat Transfer*, Vol. 5, No. 2, pp. 157–165, Apr.-Jun. 1991.

[8] Park, C.S., "Studies of Radiative Emission from the Simulated Shock Layer of the Huygens Probe" *Ph.D. Thesis*, Stanford U., 1991.

[9] Noca, M., Bailey, R., and Dyke, R., "Titan Explorer Mission Trades from the Perspective of Aerocapture," AIAA Paper No. 2003-4801, Jul. 2003.

[10] Haw, R., "Approach Navigation for a Titan Aerocapture Orbiter," AIAA Paper No. 2003-4802, Jul. 2003.

[11] Justus, C. and Duvall, A., "Engineering-Level Model Atmospheres for Titan and Neptune," AIAA Paper No. 2003-4803, Jul. 2003.

[12] Masciarelli, J. and Queen, E., "Guidance Algorithms for Aerocapture at Titan," AIAA Paper No. 2003-4804, Jul. 2003.

[13] Way, D., Powell, R., Masciarelli, J., and Starr, B., "Aerocapture Simulation and Performance for the Titan Explorer Mission," AIAA Paper No. 2003-4951, Jul. 2003.

[14] Laub, B., "Thermal Protection Concepts and Issues for Aerocapture at Titan," AIAA Paper No. 2003-4954, Jul. 2003.

[15] Hrinda, G., "Structural Design of the Titan Aerocapture Mission," AIAA Paper No. 2003-4955, Jul. 2003.

[16] Dyke, R., "Planetary Probe Mass Estimation Tool Development and Its Application to Titan Aerocapture," AIAA Paper No. 2003-4956, Jul. 2003.

[17] Takashima, N., Hollis, B., Olejniczak, J., Wright, M., and Sutton, K., "Preliminary Aerothermodynamics of Titan Aerocapture Aeroshell," AIAA Paper No. 2003-4952, Jul. 2003.

[18] Olynick, D., Chen, Y.K., Tauber, M.E., "Aerothermodynamics of the Stardust Sample Return Capsule," *J. of Spacecraft and Rockets*, Vol. 36, No. 3, May-June 1999, p. 442-461.

[19] Hartung, L.C., Mitcheltree, R.A., and Gnoffo, P.A., "Coupled Radiation Effects in Thermochemical Nonequilibrium Shock-Capturing Flowfield Calculations," *J. of Thermophysics and Heat Transfer*, Vol. 8, No. 2, Apr.-Jun. 1994, p. 244-250.

[20] Wright, M.J., G.V. Candler, and D. Bose, "Data-Parallel Line Relaxation Method for the Navier-Stokes Equations," *AIAA Journal*, Vol. 36, No. 9, pp. 1603-1609, Sep. 1998.

[21] Gnoffo, P.A., "An Upwind-Biased Point-Implicit Relaxation Algorithm for Viscous, Compressible Perfect-Gas Flows," NASA TP-2953, Feb. 1990.

[22] Park, C., *Nonequilibrium Hypersonic Aerothermodynamics*, Wiley & Sons, 1990.

[23] Whiting, E.E., Yen, L., Arnold, J.O., and Paterson, J.A., "NEQAIR96, Nonequilibrium and Equilibrium Radiative Transport and Spectra Program: User's Manual," NASA RP-1389, Dec. 1996.

[24] Hartung-Chambers, L., "Predicting Radiative Heat Transfer in Thermochemical Nonequilibrium Flow Fields, Theory and User's Manual for the LORAN code," NASA TM-4564, Sep. 1990.

[25] Reid, R.C., Jr., Rochelle, W.C., and Milhoan, J.D., "Radiative Heating to the Apollo Command Module: Engineering Predictions and Flight Measurements," NASA Tm X-58091, April 1972.

[26] Park, C., "Stagnation-Point Radiation for Apollo 4," AIAA Paper No. 2001-3070, Jun. 2001.

[27] Laux, C.O., "Optical Diagnostics and Radiative Emission of Air Plasmas," HTGL Report No. T-288, Stanford University, Aug. 1993.

[28] NIST, Washington, DC, "Atomic Spectra Database," (http://physics.nist.gov/cgi-bin/AtData/main_asd), 2003.

[29] Anderson, J.D., Jr., *Hypersonic and High Temperature Gas Dynamics,* McGraw-Hill, 1989.

THERMAL PROTECTION CONCEPTS AND ISSUES FOR AEROCAPTURE AT TITAN

Bernard Laub
NASA Ames Research Center
Moffett Field, CA 94035

ABSTRACT

A study to develop a conceptual design for an aerocapture mission at Titan was conducted by a NASA systems analysis team comprised of technical experts from several of the NASA centers. Multidisciplinary analyses demonstrated that aerocapture could be accomplished at Titan with a blunt 70° (half angle) rigid aeroshell entering the Titan atmosphere at an inertial entry velocity of ≈ 6.5 km/s. Aerothermal analyses demonstrated that the peak convective heating rates are relatively mild but the radiative heating rates, due to shock layer radiation from CN, are significantly larger and lie totally in the narrow UV band from 3500 to 4200 Å. TPS sizing analyses were conducted for a broad range of candidate TPS materials and, as expected, low density materials are the most attractive from a TPS mass standpoint. However, there is significant uncertainty associated with the interaction of low-density TPS materials with UV radiation (i.e., the potential for in-depth absorption). Consequently, the preliminary conceptual design adopted a higher mass TPS solution until the performance of low-density TPS materials can be investigated.

INTRODUCTION

Aerocapture is a flight maneuver executed by a spacecraft upon arrival at a planet in which atmospheric drag is used to decelerate the spacecraft into orbit during one atmospheric pass. This contrasts with the conventional alternatives of propulsive orbit insertion directly into the desired orbit or propulsive insertion into a large elliptical orbit followed by a long period of aerobraking to reduce the apoapsis altitude. The elements of aerocapture to establish a circular science orbit are illustrated in Figure 1.

Although aerocapture has never been attempted in any mission to date, it has long been recognized that this maneuver can greatly reduce the amount of propellant carried by the spacecraft, thereby enabling either larger payload mass fractions or smaller launch vehicles from

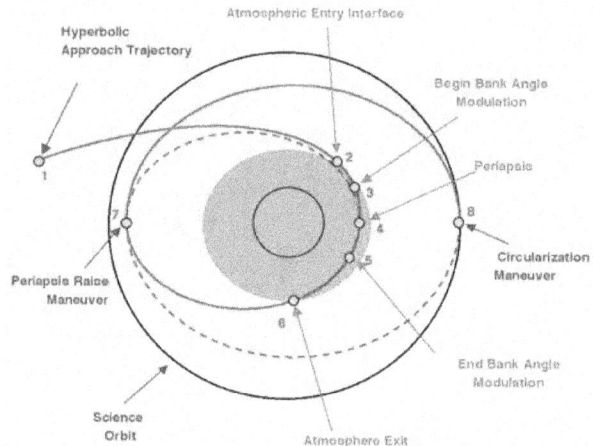

Fig. 1. The use of aerocapture to establish a circular science orbit about a planetary body[1]

Earth. The propellant mass savings of aerocapture become especially significant for missions requiring large velocity changes for orbit insertion, either because a low circular orbit is required or because the approach velocity is high. In this context, several missions in NASA's Space Science Strategic Plan are either enabled or greatly enhanced by the propellant savings afforded by aerocapture technology. These missions include Mars Sample Return, Titan Explorer, Neptune Orbiter and Venus Surface Sample Return. Other severely mass-constrained planetary missions in the Discovery Program and any secondary payload programs would also clearly benefit from the availability of aerocapture technology.

Through detailed trade studies, the mission analysis team determined that aerocapture at Titan could be accomplished with a rigid aeroshell configured as a 70 degree half-angle blunt cone forebody with a lift-to-drag (L/D) ratio of 0.25 and a ballistic coefficient (M/C_DA) of ≈ 90 kg/m². The configuration is illustrated in Figure 2. Extensive mission analysis studies[3] determined that a viable Titan mission, delivering an orbiter with a mass of 590 kg, could be launched around December 2010 and, with use of an Earth Gravity Assist (EGA) and Solar Electric Propulsion (SEP), flight time to Titan could be reduced to 5.9 years with an inertial entry velocity of ≈ 6.5 km/s at an altitude of 1000 km.

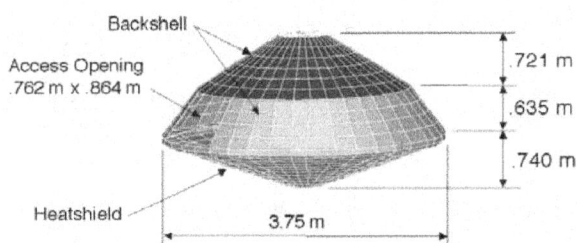

Fig. 2. Aeroshell configuration for Titan aerocapture[2]

TITAN ATMOSPHERE

The atmosphere around Titan is composed primarily of nitrogen with some argon and methane. There is some uncertainty about the concentrations of argon and methane, which leads to uncertainties in the density distribution through the atmosphere. Yelle[4] developed engineering models for atmospheric density shown in Figure 3.

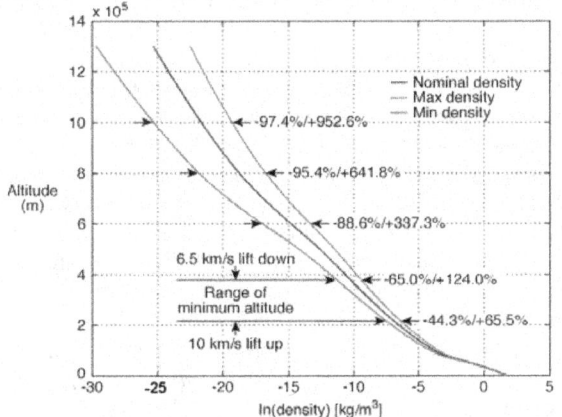

Fig. 3 Yelle engineering models for the density of the Titan atmosphere[†]

The variations in density are attributed to different models for methane concentration. The figure indicates the range of density variation of the minimum and maximum density models relative to the nominal model at selected altitudes. The figure also shows the minimum altitude range for candidate aerocapture trajectories. In these models the methane concentration is inversely proportional to mean density. As shown in Figure 4, the molar percent CH_4 is nearly constant over the altitude range where the energy of 6.5 km/s and 10

[†] The term "lift down" employed in the figure is associated with the overshoot trajectory and indicates that the lift vector is in the trajectory plane and is always pointing downward during the entire aerocapture trajectory. Conversely, "lift up" is associated with the undershoot trajectory and indicates that the lift vector is always pointing up.

km/s aerocapture entries would be dissipated. Furthermore, the composition does not vary with density perturbations. The minimum density atmosphere's methane content drops 50% between the surface and 6 km due to CH_4 condensation. However, CH_4 concentrations in the nominal and maximum density atmospheres are below the saturation level and thus remain constant to the surface.

AEROCAPTURE TRAJECTORIES

Starr & Powell[5] evaluated potential aerocapture trajectories at Titan with consideration of a range of ballistic coefficients and the uncertainties in atmospheric density. Guidance, navigation and control (GN&C) were limited to controlling the lift vector through bank angle modulation. The limiting cases are undershoot trajectories, where the lift vector is always pointing up, and overshoot trajectories, where the lift vector is always pointing down. Four trajectories,

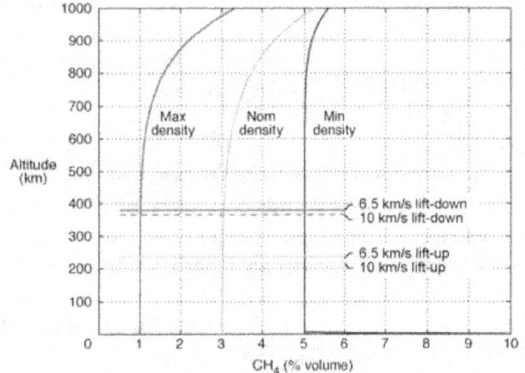

Fig 4. Models for methane concentration in Titan atmosphere

shown in Figure 5, were selected that bound the limits of peak heating rate and maximum total heat load

CONVECTIVE AND RADIATIVE HEATING

Convective and radiative heating rates[‡] at the stagnation point were calculated[6] for the aeroshell configuration described previously and the four limiting trajectories presented in Figure 5. Stagnation point convective heating was calculated using the Fay-Riddell correlation[7] and later confirmed with axisymmetric Computational Fluid Dynamics (CFD) solutions performed with the DPLR code.[8] Non-equilibrium radiation calculations were performed with the NEQAIR code.[9] As shown in Figure 6, the peak

---

[‡] Without consideration of the effects of ablation

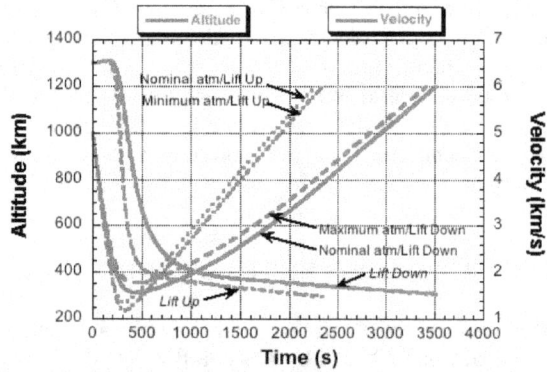

Fig. 5. Limiting aerocapture flight trajectories

stagnation point convective heating rates are less than 50 W/cm² and the undershoot trajectories (lift up) result in higher peak heating rates in comparison to the overshoot (lift down) trajectories. However, the peak stagnation point radiative heating rates are substantially larger. For the undershoot trajectories (lift up), peak stagnation point radiative heating rates are in the 120-150 W/cm² range. For the overshoot trajectories (lift down), peak stagnation point radiative heating rates are in the 45-85 W/cm² range.

It is worth noting that the convective heating is relatively insensitive to the concentration of methane assumed in the atmosphere. However, the radiative heating is very sensitive to methane concentration as all of the radiative heating is from CN formed in the shock layer from the interaction of dissociated methane with nitrogen. Consequently, the higher the methane concentration in the atmospheric model is (see Fig. 4), the higher are the radiative heating rates. While the non-equilibrium radiation environment was calculated with the most up-to-date chemistry and radiation models available, it should be recognized that there is significant uncertainty associated with these predictions. Shock tunnel tests at NASA Ames are planned within the next year to measure CN radiation at relevant conditions to enable validation and/or update of existing radiation heating models for Titan entry.

It must also be noted that the overshoot trajectories spend a longer time in the atmosphere than the undershoot trajectories. Consequently, the total convective heat load is typically larger for the overshoot trajectories in comparison to the undershoot trajectories. That is not necessarily true for the radiative heating due to the significant differences in the heat flux levels predicted for the range of trajectories and atmospheric models considered. This is demonstrated in Table 1, which summarizes the total convective and

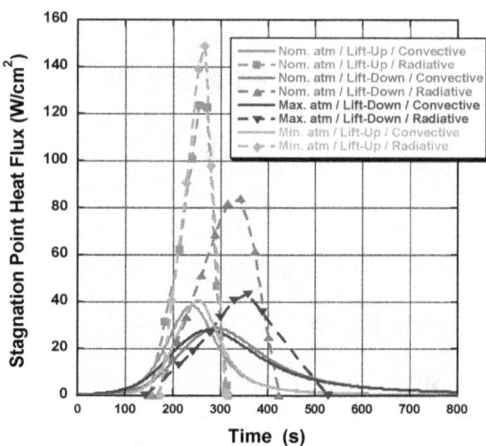

Fig. 6. Stag point heating for limiting trajectories

radiative heat loads for the four limiting trajectories considered. Note the similarities in the convective heat loads for the two overshoot and two undershoot trajectories. However, the radiative heat loads exhibit a dependence on methane concentration, that is, the minimum density atmosphere model has the highest methane concentration (see Fig. 4) while the maximum density atmosphere model has the lowest methane concentration. Table 1 suggests that total radiative heat load is dependent on time in the atmosphere (for example, compare the values for the lift up and lift down trajectories for the nominal density atmosphere) but a stronger dependence on the atmospheric model (e.g., compare the values for the maximum atmosphere on a lift up (undershoot) trajectory with the values for the minimum atmosphere on a lift down (overshoot) trajectory).

Table 1. Total heat loads for limiting aerocapture trajectories

Atmosphere model/ aerocapture trajectory	Convective heat load (J/cm²)	Radiative heat load (J/cm²)
Nominal atm/Lift up	5,500	10,021
Nominal atm/Lift down	7,500	12,090
Maximum atm/Lift down	7,700	8,393
Minimum atm/Lift up	5,200	15,769

CANDIDATE TPS MATERIALS

Given the range of convective and radiative heating described above, a range of thermal protection materials was identified as candidates for Titan aerocapture applications. The candidate TPS materials are summarized in Table 2.

Table 2. Candidate forebody TPS materials for Titan aerocapture[§]

Material	Shuttle tiles	SLA-561V	SRAM 14 (ARA)	SRAM 17 (ARA)
Density (g/cm³)	0.192-0.352	0.256	0.224	0.272
Description	Low-density, glass-based ceramic tile with glass-based coating.	Low-density cork silicone composite in Flexcore honeycomb.	Low-density cork silicone composite fabricated w/strip collar bonding technique.	Low-density cork silicone composite fabricated w/strip collar bonding technique.
Optical properties Solar absorptance Total hemis. emittance		- 0.50 - 0.78	- 0.50 - 0.78	- 0.50 - 0.78
Performance limits	44 W/cm² (certified)	Ablative. Char spall at p > 0.25 atm. No (little) recession at $\dot{q} < 100$ W/cm².	Ablative. No (little) recession at $\dot{q} < 100$ W/cm². Differential recession between composite ad interface strips may cause boundary layer transition. Other limits (if any) unknown.	Ablative No (little) recession at $\dot{q} < 100$ W/cm² Differential recession between composite ad interface strips may cause boundary layer transition. Other limits (if any) unknown.
Uncertainties	May not be adequate for current estimates of peak convective heating. Will probably be transparent to CN radiation; a show-stopper	Low density, porous material. In-depth radiant transmission may be important.	Low density, porous material. In-depth radiant transmission may be important.	Low density, porous material. In-depth radiant transmission may be important.

The list encompasses well-characterized materials such as the Shuttle tiles, SLA-561V (used on Mars Viking, Mars Pathfinder, and Mars Exploration Rover) and fully dense carbon phenolic (used on the Pioneer Venus and Galileo entry probes). It also includes some more recent TPS materials such as PICA (used as the forebody TPS on Stardust), SIRCA (used as the Backshell Interface Plate on Mars Pathfinder and Mars Exploration Rover), and the Genesis Concept (employed by Lockheed-Martin as the forebody TPS on

---

[§] The list of candidate materials was assembled based on predictions of convective heating for Titan aerocapture but prior to the availability of predicted radiative heating. The magnitude of the predicted peak radiative heating rates would eliminate some of the candidate materials from serious consideration.

the Genesis spacecraft). The list has some developmental materials such as low-density silicone and phenolic-based composites being developed by Applied Research Associates (ARA), TUFROC, a multilayer, non-ablative concept being developed by NASA Ames Research Center, and Acusil I, a moderate density silicone in honeycomb developed by ITT Aerotherm.

Most of these materials are organic resin-based composites that will pyrolyze when heated leaving a carbonaceous char at the surface. Table 2 provides the virgin density of these materials, a general description of their composition and construction, important optical properties (where known or can be estimated), some comments on their performance limits (where known or can be estimated), and some comments on performance uncertainties as perceived by the author.

Material	SIRCA	PICA	TUFROC	SRAM 20 (ARA)
Density (g/cm^3)	0.192-0.352	0.256	Varies with layer sizing	0.32
Description	Low-density ceramic tile impregnated with silicone resin.	Low-density carbon fiberform partially filled with phenolic resin.	Multilayer composite: carbon fiberform/AETB tile with high emissivity, high temperature surface treatment	Low-density cork silicone composite fabricated w/strip collar bonding technique.
Optical properties Solar absorptance Total hemis. emittance	- 0.28 - 0.92	- 0.90 - 0.80	- 0.90 - 0.90	- 0.50 - 0.78
Performance limits	Ablative. No (little) recession at $\dot{q}$ < 100 W/cm^2. Rapid recession (melting) at higher heat fluxes. Fabricated as tiles.	Ablative. No recession in non-oxidizing atmosphere at $\dot{q}$ < 1000 W/cm^2. Excellent low density ablator, but not best insulator.	Non-ablative. May be usable to $\dot{q}$ ~ 300 W/cm^2. Has to be fabricated as tiles.	Ablative. No (little) recession at $\dot{q}$ < 100 W/cm^2. Differential recession between composite ad interface strips may cause boundary layer transition. Other limits (if any) unknown.
Uncertainties	Low-density, glass-based material. In-depth radiant transmission may be important.	Low density, porous material. In-depth radiant transmission may be important.	Heat flux limit currently uncertain. Spectral emittance data on coating shows very high emittance at CN violet wavelengths. Will absorb (not transmit) UV radiation.	Low density, porous material. In-depth radiant transmission may be important.

Table 2. Candidate forebody TPS materials for Titan aerocapture (continued)

From the TPS standpoint, total heat loads for aerocapture are typically much larger than for direct entry. Consequently, the lowest mass TPS solutions will be good insulators, a characteristic that is usually associated with low-density materials. However insulation performance must be balanced with ablation performance, for example, too much surface recession can lead to alterations in shape that can affect aerodynamic performance. High-density materials are usually employed to minimize surface recession. However, the peak heating rates predicted for Titan aerocapture are within the capabilities of several of the low-density candidate materials, that is, they can provide good insulation performance with minimal surface recession. Consequently, they are considered the most attractive candidate materials from the standpoint of minimizing TPS mass.

TPS MASS ESTIMATES

To provide an estimate of TPS mass, preliminary forebody TPS sizing analyses were done for the four limiting aerocapture trajectories shown in Figure 5 for many (but not all) of the candidate materials listed in Table 2. The analyses were limited to the stagnation point heating shown in Figure 6 and, for purposes of estimating TPS mass, it was assumed that the nominal (without margin) stagnation point TPS thickness is

Table 2. Candidate forebody TPS materials for Titan aerocapture (concluded)

Material	PhenCarb-20 (ARA)	Acusil I	Carbon phenolic	Genesis concept
Density (g/cm³)	0.32	0.48	1.45	Varies with layer sizing
Description	Low-moderate density phenolic composite fabricated w/strip collar bonding technique.	Moderate density filled silicone in Flexcore honeycomb.	Fully dense tape-wrapped or chopped molded heritage material.	Carbon-carbon facesheet over carbon fiberform insulator.
Optical properties				
Solar absorptance	- 0.50	0.54	- 0.90	- 0.90
Total hemis. emittance	- 0.78	0.95	- 0.90	- 0.90
Performance limits	Ablative. No recession in non-oxidizing environment. Differential recession between composite ad interface strips may cause boundary layer transition. Other limits (if any) unknown.	Ablative. No recession (in air) at $\dot{q} < 100$ W/cm². Will recede in air at higher heat fluxes (assumed by oxidation).	Ablative. No recession expected for Titan aerocapture environment.	Ablative. No recession expected for Titan aerocapture environment.
Uncertainties	Higher density and higher char yield of phenolics may mitigate in-depth radiant absorption at the penalty of higher thermal conductivity.	Higher density and good char yield may mitigate in-depth radiant absorption. Not as good an insulator as lower density materials. Difficult to fabricate in thicknesses > 2 inches.	High density and high char yield will mitigate in-depth radiant absorption. Not a very good insulator. Without recession, will be very heavy (a poor choice for this application).	High-density carbon facesheet will mitigate in-depth radiant absorption. Carbon fiberform not the best insulator, but better than fully dense carbon phenolic.

applied uniformly on the forebody. It was also assumed that the TPS is adhesively bonded to a rigid substructure consisting of 0.0376 mm thick graphite polyimide facesheets (front and back) on a 31.75 mm thick aluminum honeycomb. The density for the honeycomb and graphite polyimide facesheets is 0.069 g/cm³ and 1.0 g/cm³, respectively. It was assumed that all materials are at a uniform temperature of –74.8°C at atmospheric interface. It was also assumed that all candidate materials absorb CN radiation at the surface and perform as thermochemical ablators, that is, no spall or melt runoff. Analyses were performed to determine the thickness required for each candidate material to limit the maximum bondline temperature to 250°C. It should be noted that different individuals did the analyses for different materials, but all used comparable analysis tools that address the fundamental physical and chemical mechanisms associated with the thermal/ablation performance of these materials in the Titan atmosphere. Some of the materials models are very mature and have been validated with extensive laboratory and arc jet test data. In contrast, some of the other materials are relatively new and their models are based on limited laboratory and arc jet test data.

The results of the analyses demonstrated that, for the four trajectories considered, the TPS thickness is significantly larger for the overshoot trajectories (lift down) than for the undershoot trajectories (lift up). This was the case for all of the candidate materials considered. Furthermore, maximum bondline temperature is attained during heat soak, that is, after the end of aerodynamic heating.

Table 3. Preliminary forebody TPS sizing for Titan aerocapture

Candidate TPS Material	Maximum atmosphere - Lift Down Convective Heat Load = 7,700 J/cm^2 Radiative Heat Load = 8,393 J/cm^2		Nominal atmosphere – Lift Down Convective Heat Load = 7,500 J/cm^2 Radiative Heat Load = 12,090 J/cm^2	
	Thickness (cm)	Areal weight (g/cm^2)	Thickness (cm)	Areal weight (g/cm^2)
SLA-561V	2.44	0.626	2.43	0.622
SRAM 14	1.57	0.353	1.55	0.348
SRAM 17	1.93	0.526	1.93	0.526
SRAM 20	2.08	0.667	2.08	0.667
PhenCarb-20	2.29	0.696	2.34	0.711
TUFROC	4.88	1.117	5.13	1.181
Genesis	---	---	5.51	1.298
PICA	5.94	1.591	5.82	1.557
Carbon phenolic	8.70	13.084	8.76	13.167

The results of the TPS sizing analyses are summarized in Table 3 which only shows the thickness and areal weight requirements for the overshoot trajectories because, as stated previously, they exceeded the requirements for the undershoot trajectories.

Table 3 illustrates that the thickness and areal weight for any candidate TPS material is similar for the two overshoot trajectories despite the significant difference in the total radiative heat load. Examination of Figure 6 will illustrate that the predicted radiative heating for the nominal atmosphere model results in higher peak heat fluxes in comparison to the maximum atmosphere model, but for a shorter period of time. As it was assumed that all of these materials are surface absorbers, the higher heat fluxes result in higher surface temperatures. As surface re-radiation is the predominant energy accommodation mechanism for all of these candidate materials, they are all more "efficient" at higher heat fluxes as long as they do not experience significant surface recession.

Table 3 also illustrates that, in general, TPS areal weight increases with material density. This is clearly evident by comparing the areal weight requirements between the candidates in the SRAM family where the composition and construction are similar but the virgin densities are different. Also, note that the areal weight of PhenCarb 20 is similar to SRAM 20. However, that is coincidental; the materials have the same virgin density but different compositions, as SRAM uses a silicone resin binder and PhenCarb, a phenolic resin binder. The thermal conductivity of the PhenCarb family should be higher than that of the SRAM family, but that is compensated for by the smaller surface recession experienced by the PhenCarb in comparison to the SRAM.

This tradeoff is clearly demonstrated when one compares the areal weight of PICA with SRAM17 where the virgin densities are similar. PICA, composed primarily of carbon, has a much higher thermal conductivity than SRAM17 and this is reflected in the areal weights, with the areal weight of PICA approximately three times that of SRAM17. The fully dense carbon phenolic result clearly makes the point because, from the standpoint of composition, it is similar to PICA. Yet the almost six-fold increase in virgin density is accompanied by a significant increase in thermal conductivity, which is reflected in the areal weight requirements. The key point is that the materials that are primarily carbonaceous are excellent "ablators" but not good insulators. They are best used in environments with very high heating rates because they are most efficient when they ablate. The heating rates for Titan aerocapture trajectories are not sufficiently severe to allow these materials to perform efficiently. The low density, low thermal conductivity materials are better choices.

The Genesis and TUFROC candidates need to be viewed differently. The Genesis TPS employs a thin carbon-carbon facesheet over a carbon fiberform insulator. The carbon-carbon is robust and will not experience any recession for the range of Titan aerocapture heating environments. Although the carbon fiberform is not the best insulator, it is one of the few materials that can be employed in contact with the carbon-carbon because such a material must be stable at the surface temperatures that the C-C facesheet will attain. The TUFROC concept is different in some aspects. It is comprised of a carbon fiberform insulator bonded to an AETB low-density ceramic insulator. The two-layer composite has a high temperature, high emissivity, low catalytic efficiency surface treatment. The assembly is intended to be non-ablative, but the upper limit of the surface treatment is yet to be demonstrated. The sizing of TUFROC, a two-step process as defined by the carbon fiberform thickness, is governed by the upper temperature limit of the AETB ceramic. Once that is defined, the AETB thickness is sized to limit bondline temperature.

As a consequence of the carbon fiberform minimum thickness requirement for TUFROC, the areal weight of this concept is larger than for most of the other candidate materials. Existing data suggest that the coating can handle the range of heat fluxes anticipated for the range of Titan aerocapture trajectories considered. Furthermore, existing optical properties data on the coating indicate that it will absorb UV radiation. The results for the Genesis concept indicate that its areal weight will be slightly larger than that for TUFROC, primarily because the carbon fiberform is not as good an insulator as a composite of carbon fiberform over a low-density AETB ceramic.

TPS PERFORMANCE UNCERTAINTIES

Of some concern is the interaction of CN radiation with low density, porous TPS materials. Figure 7 illustrates the spectral distribution of the predicted[¶] CN radiation where it is seen that almost all the radiation lies in a relatively narrow band in the ultraviolet (UV) with the peak at ≈ 3800Å ($0.38\mu m$). Studies conducted during the 1980s, evaluated the performance of dozens of ablative materials exposed to high-energy lasers. The types of materials evaluated spanned the range from low-density organic resin composites to fully dense carbon-carbon composites. Materials were tested with both continuous wave (CW) and repetitively pulsed (RP) lasers at wavelengths from the visible ($0.53\mu m$) to the infrared ($10.6\mu m$). While material performance was strongly dependent on the type of material and the irradiance (heat flux) it was exposed to, the data also suggested a general trend where material performance degraded at the shorter wavelengths. Further studies demonstrated that the materials did not become semi-transparent at the shorter wavelengths, but rather the absorption length became larger as the wavelength got shorter. It should be noted that none of the materials that are TPS candidates for Titan aerocapture were evaluated under these laser studies.

_____

[¶] Nonequilibrium NEQAIR calculations

Based on the trend observed in the laser studies, the performance of the most attractive TPS candidates for Titan aerocapture is, at best, uncertain. If these low-density materials have significant absorption lengths to radiation at UV wavelengths, there is the potential that the materials may spall (blow off the surface layer due to in-depth pyrolysis). It can be argued that when these materials pyrolyze they will form a carbonaceous

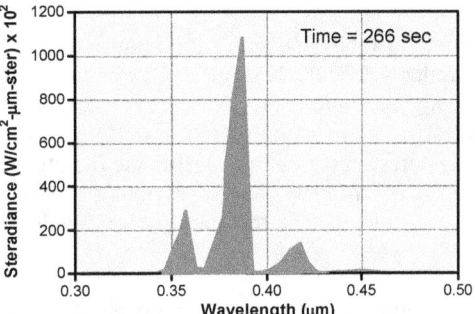

Fig. 7 Spectral distribution of CN radiation

surface char with different absorption characteristics than the virgin material; the char will be a surface absorber and/or scatter the incident radiation. That is certainly possible and, in fact, anticipated if the char has sufficient time to form and stays attached. That is the major issue because it is possible that in-depth absorption leading to periodic spall would continually remove whatever surface char is in the process of being formed.

It is also possible that pyrolysis gases injected into the boundary layer could scatter the incoming CN radiation or absorb it and re-radiate at longer wavelengths. Again, that is possible but it is very difficult to demonstrate, even analytically, due to uncertainties in the composition of the pyrolysis products and their absorption characteristics.

The only way to evaluate the performance of these materials when exposed to UV radiation is through experiments. Unfortunately, a good experimental simulation is not simple. The ideal experiment would expose these candidate materials to a radiation source at the wavelengths of interest and at relevant heat fluxes. All of the high-energy CW lasers operate in the infrared. There are excimer lasers that operate at the wavelengths of interest, but they are pulsed lasers that produce extremely high heat fluxes over a very short duration. There are some CW Argon-ion and Krypton lasers that produce radiation in the range of interest, but they are very low power.

It has been suggested that the materials be tested with a solar source. Examination of the black body spectral distribution of solar radiation illustrates that only a small fraction (<15%) of the energy lies in the UV band of interest with a significant fraction in the IR. Furthermore, the atmosphere filters out much of the UV radiation. Testing with a solar source would expose the materials to broadband radiation and the potential for in-depth absorption of UV radiation could be masked by char formation promoted by surface absorption of IR radiation.

Tests could also be done with a high power xenon lamp that simulates a 5800 K source. Examination of the black body spectral distribution from such a source illustrates that only a small fraction ($\approx$15%) lies in the band of interest. It is possible to filter the radiation from this lamp to allow only the UV radiation to reach the material sample, but at the sacrifice of much of the power.

The best option is the use of a commercially available Mercury-Xenon Lamp, which will allow the testing of 2.54 cm (1 inch) square specimens at the required energy flux levels. Mercury-Xenon short arc lamps are dosed with an exact amount of mercury and xenon gas. Due to the high operating pressure of mercury-xenon, the spectrum shows only traces from the broad, visible and infrared dominated xenon gas spectrum. The four main mercury lines dominate the spectrum, as shown in Figure 8. The first three lines provide an excellent simulation of the predicted CN radiation at Titan. The visible lines between 500 and 600 nm are easily filtered out, and can be included or not as desired. A simple cold filter can eliminate the visible and infrared lines.

A more fundamental approach would measure the relevant optical properties of the virgin (and char) candidate materials using a spectrophotometer and an integrating sphere. Such data would be very valuable but would require making (some) measurements on very thin samples. Slicing low-density materials into thin sections ($\approx$ 1 mm) would be a challenge.

From the above discussion it should be apparent that validating adequate performance of the best TPS candidates for Titan aerocapture would not be simple due to experimental limitations. Some combination of the aforementioned tests must be done and the resultant data will need to be carefully evaluated before one can design a TPS for Titan aerocapture with confidence. If adequate performance of the most attractive candidate materials when exposed to CN radiation cannot be experimentally demonstrated, any TPS design employing these materials will have to include significant "margin" to mitigate the risk.

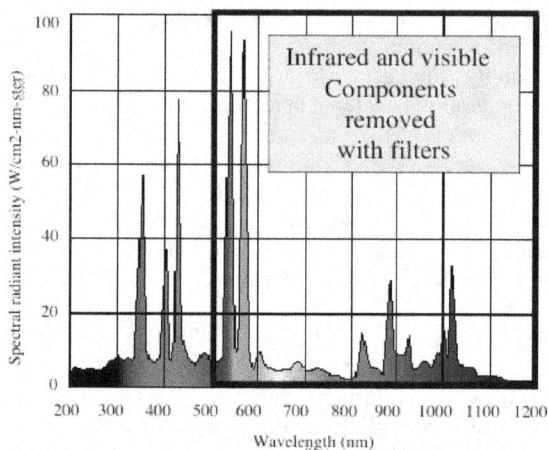

Fig. 8. Spectral output of Mercury-Xenon lamp shows good simulation of CN lines.

PRELIMINARY TPS CONCEPTUAL DESIGN

The charter of the NASA aerocapture systems analysis team is focused on identifying and prioritizing the most critical technology gaps to enable aerocapture to be utilized for Solar System Exploration (SSE) missions. The benefits of employing aerocapture for Titan Explorer can only be quantified through comparison with a design that employs chemical propulsion for orbit insertion. To enable that comparison, the aerocapture systems analysis team developed a preliminary conceptual design of an aerocapture "system" for Titan Explorer so that comparisons could be made of launch mass, payload mass, and system complexity. An estimate of TPS mass was required to support this conceptual design. For this purpose the systems analysis team preliminarily selected TUFROC for the forebody TPS. As seen in Table 3, the nominal areal weight of TUFROC (only sized at the stagnation point) is $\approx$ 1.18 g/cm^2 for the worst-case trajectory (nominal atmosphere, lift down). The surface area of the forebody aeroshell is $\approx$ 12.59 m^2, which results in a nominal forebody TPS mass of $\approx$ 149 kg. The TPS sizing studies identified several candidate materials that, potentially, would provide lower TPS mass solutions, in some cases by factors of 2-3. However, all of these attractive candidate materials need to be experimentally evaluated to ensure that potential in-depth radiant absorption will not significantly degrade material performance.

SUMMARY

A NASA systems analysis team developed a rigid aeroshell conceptual design for orbit insertion around Titan using aerocapture. Candidate TPS materials were identified and preliminary TPS sizing studies were conducted using predicted stagnation point convective and radiative heating for limiting aerocapture trajectories. The results of these analyses demonstrated that existing low-density organic resin composites provide low TPS mass solutions for this mission. However, the potential for in-depth absorption of CN radiation can degrade the performance of these materials. The interaction of these materials with UV radiation needs to be experimentally evaluated, but the options for conducting the appropriate ground test experiments simulating UV radiation at relevant heat fluxes are limited. Therefore, the preliminary conceptual design incorporated a TPS concept (TUFROC) that has been demonstrated to absorb UV radiation at the surface, but requires a significantly larger TPS mass.

ACKNOWLEDGEMENTS

The author wishes to express his appreciation to Dr. Y. -K. Chen (NASA Ames Research Center) who conducted the TPS sizing analyses for SLA-561V, PICA, and TUFROC, to Dr. Gary A. Allen (ELORET/NASA Ames Research Center) who conducted the TPS sizing analyses for fully-dense carbon phenolic, to Mr. William M. Congdon (Applied Research Associates) who conducted the TPS sizing analyses for SRAM14, SRAM17, SRAM20 and PhenCarb20, and to Mr. John A. Dec (NASA Langley Research Center) who conducted the TPS sizing analyses for the Genesis TPS concept. The author also wishes to acknowledge Dr. Donald M. Curry (NASA Johnson Space Center) and Dr. F. Neil Cheatwood (NASA Langley Research Center) who were consulted in assembling the list of candidate TPS materials for Titan aerocapture. The author also wishes to acknowledge the special contributions of Dr. Susan M. White (NASA Ames Research Center) who researched available devices to simulate the Titan radiation environment and identified the Mercury-Xenon lamp.

REFERENCES

1. Lockwood, M.K. *"Titan Aerocapture Systems Analysis Overview,"* NASA Langley Research Center, Aerocapture Systems Analysis Review, Jet Propulsion Laboratory, Pasadena, California. August 29-30, 2002.
2. Hrinda, G.A., *"Aeroshell Structure,"* NASA Langley Research Center, Aerocapture Systems Analysis Review, Jet Propulsion Laboratory, Pasadena, California. August 29-30, 2002.
3. Noca, M., *"Mission Analysis,"* Jet Propulsion Laboratory, Aerocapture Systems Analysis Review, Jet Propulsion Laboratory, Pasadena, California. August 29-30, 2002.
4. Yelle et al.: "Engineering Models for Titan's Atmosphere", in *Huygens Science, Payload and Mission*, ESA SP-1177, 1997
5. Starr, B.R. and Powell, R., *"Titan/Neptune Aerocapture System Study,"* NASA Langley Research Center, Hampton, VA, March 29, 2002.
6. Olejniczak, J. and Allen, G.A., Jr., *"Radiative Heating/TPS Sizing Results,"* NASA Ames Research Center, May 2, 2002.
7. Fay, J.A. and F.R. Riddell, *"Theory of Stagnation Point Heat Transfer in Dissociated Air,"* J. Aero. Sci., 25, (2), 73-85, 1958.
8. Wright, M.J., G.V. Candler, and D. Bose, *"Data-Parallel Line Relaxation Method for the Navier-Stokes Equations,"* AIAA Journal, Vol. 36, No. 9, pp. 1603-1609, Sep. 1998.
9. Whiting, E.E., Park, C., Liu, Y., Arnold, J.O., and Paterson, J.A., *"NEQAIR96, Nonequilibrium and Equilibrium Radiative Transport and Spectra Program: User's Manual,"* NASA RP-1389, Dec. 1996

STRUCTURAL DESIGN OF THE TITAN AEROCAPTURE MISSION

Glenn A. Hrinda
NASA Langley Research Center, Hampton VA

ABSTRACT

A major goal of NASA's planetary exploration efforts is to create affordable spacecraft capable of delivering science experiments for long duration periods. To help achieve this goal the aerocapture technique for slowing a spacecraft has been investigated and appears to produce less vehicle mass then an all-propulsive mission. A conceptual spacecraft was designed and studied for an aerocapture mission to Titan, Saturn's largest moon. The spacecraft is an Orbiter/Lander combination that separates prior to aerocapture at Titan. The structural challenges faced in the design will be discussed as well as optimization sizing techniques used in the Orbiter's aeroshell structure. Design trades required to optimize the structural mass will be presented. Member sizes, concepts and material selections will be presented with descriptions of load cases and spacecraft structural configurations. Areas of concern will be highlighted for further investigation. This study involved the colaberation efforts of NASA representatives from Langley Research Center (LARC), Jet Propulsion Lab (JPL) and Ames Research Center (ARC). The concept design borrowed from existing flight hardware as much as possible.

INTRODUCTION

The structural sizing for a conceptual aerocapture spacecraft to Titan was required to obtain mass estimates based on current sizing methods. Finite element analysis (FEA) and HyperSizer™ sizing software was used to model the launch stack assemble that included the Propulsion Module (PM), the Orbiter and the Lander. The Orbiter spacecraft performs aerocapture at Titan and is designed to withstand atmospheric heating. The Lander is a sphere-cone and was considered as a concentrated mass. No aeroshell design analysis was performed on the Lander. The launch vehicle used was a Boeing Delta IV heavy with a 4 meter fairing. The spacecraft integration into a Delta IV heavy launch vehicle was achieved through trade studies focusing on mission performance necessary for an aerocapture mission. Three primary design objectives were: minimum structural mass, dynamic modes at launch were meet or exceeded and stress levels were within margin with minimal deflections. Load cases and frequency minimums at launch came from the Boeing Payload Planners Guide. Maximum loading at launch and during entry at Titan was used to design the spacecraft structure. The lowest predicted natural dynamic modes were investigated to identify any low frequency problems with the spacecraft.

The structural design used composites for the Orbiter aeroshell and a truss system to join the stack components. Modeling efforts were kept as simple as possible to shorten modifications occurring as the design progressed. HyperSizer™ sizing software was found beneficial in sizing the Orbiter's aeroshell. The software's ability to optimize composite sections without refining mesh densities and geometry was demonstrated throughout the design's progress.

NOMENCLATURE/ABBREVIATIONS

CG	Center of Gravity
FEA	Finite Element Analysis
FEM	Finite Element Model
HGA	High Gain Antenna
NSM	Non-structural mass
PM	Propulsion Module
TPS	Thermal Protection System

CONCEPTUAL TITAN DESIGN

The Titan aerocapture spacecraft is a stacked configuration requiring a three component stack consisting of a Propulsion Module (PM), an Orbiter and Lander. The three separate spacecraft were combined to form a launch stack capable of fitting into a 4 meter Delta IV fairing. Each vehicle must be able to separate during the Titan aerocapture mission sequence. The PM provides all thrust maneuvering to get the Orbiter and Lander near Titan. The Orbiter and Lander will then separate from the PM and then from each other. The Lander descends to the Titan surface in a Huygens type aeroshell. The Orbiter continues through the thin Titan atmosphere and begins aerocapture until achieving its mission orbit. The Orbiter was the only vehicle designed to take advantage of aerocapture. Once in orbit around Titan, the Orbiter will support an on orbit relay station for the Lander.

The conceptual spacecrafts were used as a baseline to test design and analysis methods used among the various NASA centers involved with aerocapture vehicle designs. This study focused on the weight reduction and strength requirements of the major load carrying structural members. The design attempts to maintain an axial load path direction starting with the Lander, into the Orbiter through its payload pallet and heatshield and final into the PM.

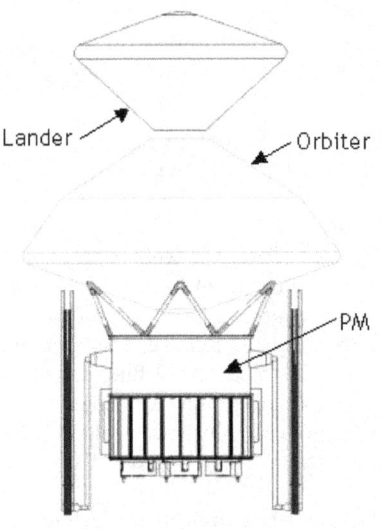

Figure 1. Launch Stack

The structural analysis performed in this study helped verify the stack arrangement and size the Orbiter aeroshell and support structure for the generic Lander. Investigating various stack arrangements showed that a truss would provide the lightest structure for supporting the Lander. A truss was also used for the PM to Orbiter adapter structure. The study used a launch load envelop for the Delta IV heavy. The Orbiter maximum diameter was set to 3.75 meters and used a heatshield cone and biconic backshell as shown in figure 1. The PM was modeled to include its stiffness contribution in determining overall stack frequency during launch. Launch loads were taken from the Boeing Delta IV Payload Designers manual (ref. 1).

STACK CONFIGURATIONS

The stacking sequence of the PM, Orbiter and Lander was decided upon after several trials of the three components arranged in different configurations. Each configuration had its abilities compared with each other until the stack shown in figure 1 was chosen. This arrangement was used after various stack sequences were attempted to find a stack able to meet strength, dynamics and center-of-gravity (CG) requirements. The diagrams in figure 2 represent a sample of the many stack sequences of the PM, Orbiter and Lander attempted during the design trade studies. The final stack configuration used in the design placed the Lander on top of the Orbiter.

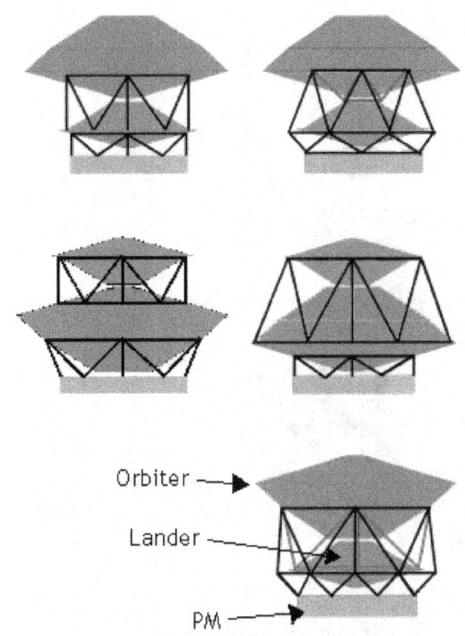

Figure 2. Trial Launch Stack Configurations

The configurations placing the Orbiter on top were dismissed because of the large mass of the Orbiter and low lateral stack bending frequency that is created during launch. The configurations produced heavy structures due to the increased stiffness requirements necessary for raising the lateral bending frequency above 10 Hz. One of the design goals was to minimize structural mass and maintain a minimum frequency of 10Hz lateral and 27Hz axial. These values were taken from the design guide in reference 3. The final configuration produced the minimal structural mass and maintained design stiffness requirements.

A generic sphere-cone Lander with a mass of 400 kg was assumed in the study and was modeled as a lumped mass with rigid connections to its outer diameter. A truss is used to create a load path from the Lander, through the Orbiter and into the PM. During the mission the Orbiter and Lander separate from the PM. The truss adapter to the PM is jettisoned with the PM and the six attachment points to the Orbiter heatshield are plugged. The method for plugging the heatshield penetrations will require further study. One possibility is to mechanically activate panels to cover the attach points. The Lander will separate from the Orbiter and head directly for the Titan surface. The upper truss supporting the Lander will then separate from the Orbiter. Aerocapture of the Orbiter will then commence at Titan.

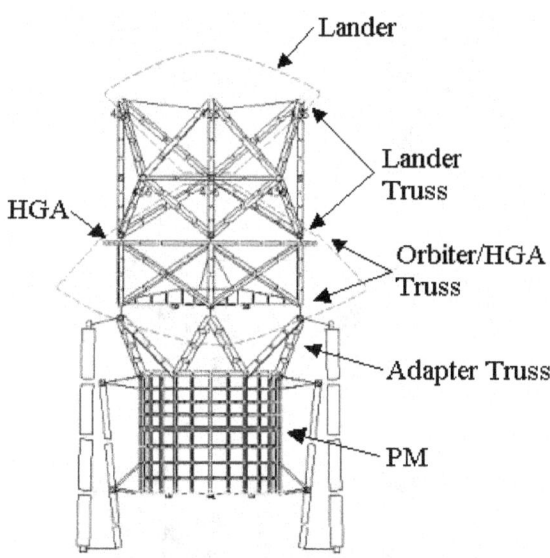

Figure 3. Stack Truss

ORBITER AEROSHELL DESIGN

The Orbiter spacecraft structure consists of the fore body heatshield, biconic backshell, cap plate and internal support structure supporting the Orbiter's payload. The largest payload components in the Orbiter are the 2.75m High Gain Antenna (HGA) and the spherical hydrazine tank. Figure 4 shows these components plus the arrangement of other internal components carried by the Orbiter. The payload deck is a hexagonal shaped aluminum honeycomb panel that extends to the aeroshell at six separation points. All payload deck items are modeled as concentrated masses with their CG offset made using rigid elements as required. The Orbiter aeroshell is supported by the adapter truss attached to the heatshield. The Lander is supported by a tube truss system that penetrates the Orbiter backshell at four locations. The load path continues straight down through an internal structure that also supports the HGA.

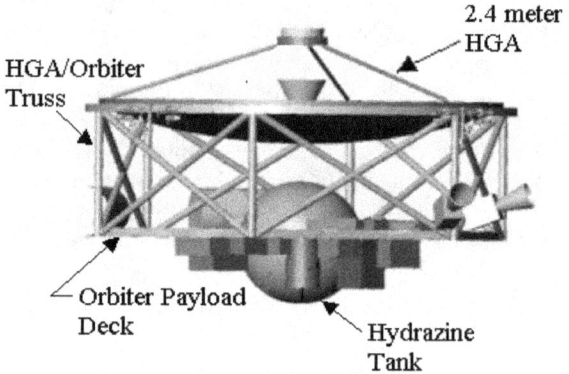

Figure 4. Orbiter HGA and payload deck (Lander not shown)

A payload pallet ring is used to transfer the loads coming from the Lander as well as the payload deck to six hard points on the Orbiter's heatshield. The six hard points are equally spaced around the perimeter of the payload ring and represent penetrations through the heatshield. The Orbiter aeroshell FEM is pinned at the six hard points. Concentrated masses were used to model the internal payloads along with rigid elements to properly locate CG's.

The load contribution from the Lander and Orbiter is carried into an adapter truss through the six points on the heat shield. The adapter truss tapers down to fit the front of the PM completing the load path. The choice of allowing the load path to continue through the heat shield raises obvious concerns with the thermal protection system (TPS) being compromised. The mass of the Orbiter was reduced by not using its aeroshell to support the Lander mass. The stack concept relied on keeping load paths running axially through a tubular space truss. The six penetrations in the heatshield were accepted in this study and referred to as a detail requiring further investigation.

The Orbiter was analyzed using a combination of nastran finite element analysis (fea) and Hypersizer™ commercial sizing and optimization software. In order to utilize Hypersizer™, a coarse grid nastran fea was created with all non-structural masses (NSM) and mission loads of interest. The major NSM contribution was the heatshield and backshell TPS. Other NSM included Orbiter payload and aeroshell separation mechanisms as well as allocations for the six attach points through the heatshield. Figure 5 shows an exploded view of the FEM used to create a HyperSizer™ model of the Orbiter's aeroshell.

<u>Orbiter Model</u>

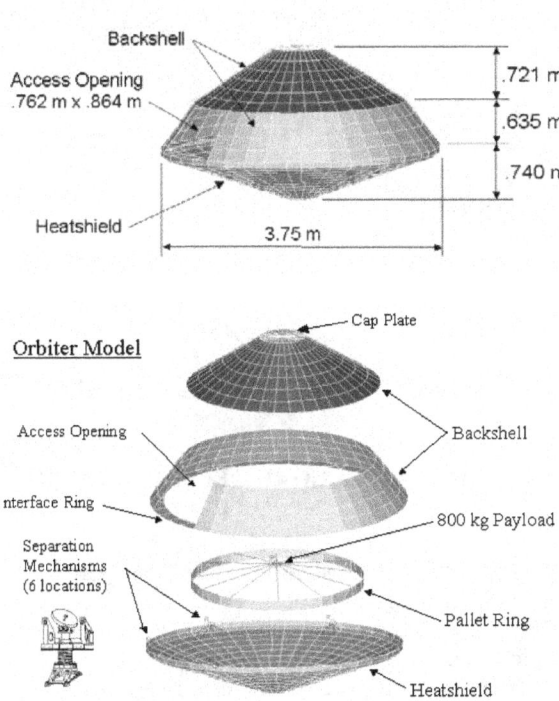

Figure 5. Orbiter Aeroshell FEM

The mesh size was kept coarse, however included enough detailing of the aeroshell geometry to accurately calculate the element forces required in Hypersizer™. An opening in the backshell was modeled to represent an access panel to the Orbiter payload. No attempt was made to stiffen the opening by modeling the door or method of attachment to the backshell. The mass of the door was treated as a NSM with smearing at the nodes. The nastran finite element model (FEM) of the Orbiter was created only with basic nastran elements: quad4, tria3, conm2 and rbe2's. These elements are easily supported by Hypersizer™ and were imported to form a Hypersizer™ model of the Orbiter aeroshell.

Several model configurations and load cases were used to find worse case conditions on the Orbiter aeroshell. The first configuration studied was the Orbiter in the launch mode with accelerations based on the Delta IV payload guide (ref. 3). The loads used were 3 g's lateral and 7 g's axial. The combined loads were the absolute maximums in the Delta IV launch load envelope. No assessment of acoustic energy and shock spectra on the total payload was attempted during the design.

The next load cases investigated were aerocapture entry loads of the Orbiter through the Titan atmosphere. Two load cases were investigated for different entry velocities. A 6.5 km/s and 10 km/s entry load cases were analyzed. The peak aero loads were obtained from CFD analysis based on the two trajectory cases. The loads were assumed to act normal to the heatshield and evenly distributed. The following loads were used:
6.5 km/s Entry loads: 4 G axial with 3146 Pa on heatshield
10 km/s Entry loads: 10.3 G axial with 8997 Pa on heatshield

The launch load forces were imported into Hypersizer™ to start sizing of the aeroshell. The Orbiter's Hypersizer™ model was divided into different components for sizing. The approach was to size the heatshield as one uniform thickness as well as the lower and upper backshell and cap plate. The optimization concepts used were: honeycomb core with face sheets, blade stiffened panels and isogrids. Each concept had dimension variables that were used to find the optimal aeroshell geometry such as: blade separation distances, core and face sheet thickness, blade depths and thickness.

The TPS non-structural mass was added inside Hypersizer™. The mass could easily be changed on one of Hypersizer's™ user input screens. This feature of the software was helpful for modifying the model to suit different TPS trial materials and thickness. The final TPS material used in the design of the Orbiter was TUFROC on the heatshield with an aerial density of 1.181 g/cm². This density was held constant over the heatshield. The backshell and cap plate used SLA with an aerial density of .187 g/cm². The density was also constant over both surfaces. The TPS masses were exported as nastran conm2's and evenly distributed at the element nodes.

WHY USE HYPERSIZER?

Spacecraft structures contain complex geometry and load distributions that are highly indeterminate and historically demanded finite element analysis (FEA) to solve. Performing structural analysis and sizing optimization has required large degree-of-freedom models with long solution run times. A software product called HyperSizer^{TM} can help simplify structural sizing and reduce design analysis time. HyperSizer^{TM} helps to automate the sizing of structures by reducing launch acceleration and entry loads into force and moment components on panels and beams throughout the spacecraft. The sizing includes finding the optimal material combinations, panel and beam dimensions such as thickness, depths and spacing. The code is not a finite element analysis or computer aided design package. HyperSizer^{TM} adds to the capabilities of these tools to allow the engineer to design, size and perform detailed failure analysis on a complete vehicle.

The Orbiter's aeroshell design was used to demonstrate the software's composite design capability and use in conceptual designs. A new mass-sizing tool is under development for planetary spacecraft at LARC. The tool will have the ability to link spreadsheet user inputs into HyperSizer^{TM} for composite structure sizing. This will greatly improve structural mass estimates and lessen analysis time usually dominated by FEM creation and modifications.

OPTIMIZATION CAPABILITIES

Optimization capabilities within HyperSizer^{TM} include finding minimum weight panel or beam concepts, material selections, cross sectional dimensions, thickness and lay-ups from a database of 50 different stiffened and sandwich panel designs as well as a database of composite, metallic, honeycomb and foam materials. The database is used to define structural families inside HyperSizer^{TM}. The structural families include definitions for panels and beams such as the "uniaxial stiffened family", the "unstiffened plate/sandwich family" and the "open beam family".

The panels shown in figure 6 below represent some of the typical families of structural panels available in HyperSizer^{TM}. The panels may be stiffened with typical aerospace shapes or corrugated. The grid-stiffened family of panels has recently been added to HyperSizer^{TM}. This allows for the sizing optimization of isogrids, orthogrids and general grid rib-stiffened panel concepts with either isotropic or composite materials

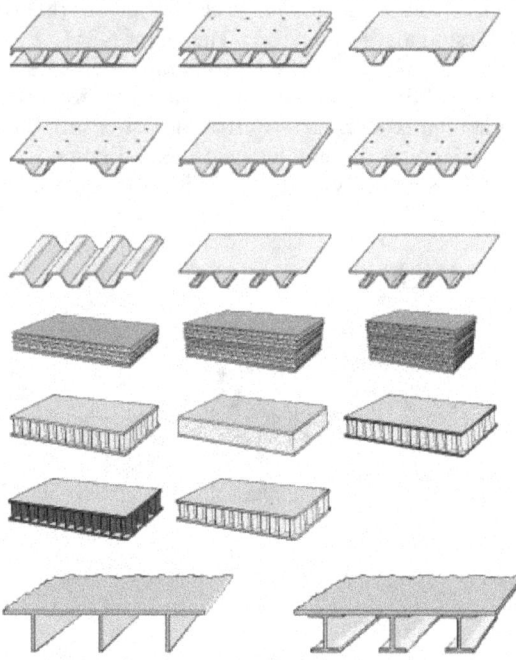

Figure 6. Typical HyperSizer^{TM} panels

ORBITER PAYLOAD DECK AND HGA SUPPORT

The Orbiter payload deck and HGA support were modeled with plate and beam elements. The payload deck was a flat hexagonal shaped plate with a large hole cutout for a hydrazine tank carried into aerocapture orbit. The six corners of the hexagon platform extend to the outer diameter of the Orbiter. The platform lies in the same plane as the backshell/heatshield separation plane. The HGA is supported by an internal truss that also connects to the Lander truss. Loads from the Lander travel through the HGA support and into the payload ring located below the payload deck. Loads are then transferred through the heatshield structure and into the Orbiter/PM adapter.

PROPULSION MODULE

The PM was modeled as shown in figure 7. The bulk of the module was made of aluminum channels and distributed lumped masses. Two solar arrays were also modeled and appear on the module sides. The solar arrays were modeled as beams having an approximate stiffness of the array panels. The modeling effort attempted to accurately capture the correct stiffness and mass of the module without fine detailing of the meshes. A concentrated mass with rigid elements was used for the propellant tank. Support structure for the tank was also provided with a truss system tying into a ring frame. A cylindrical wall stiffened with beam elements form the main thrust tube. The aft end of the tube was pinned with the forward end attaching to the Orbiter/PM adapter truss. The PM was modeled to help determine overall stack frequency at launch. By including the stiffness from the PM in the dynamic analysis, better determinations of the lowest modes were found.

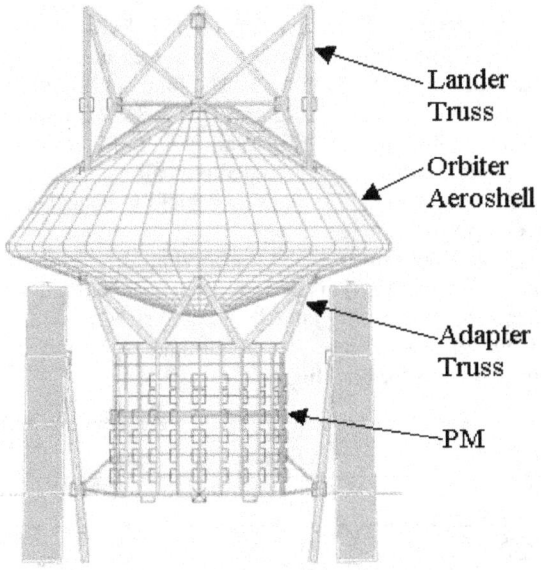

Figure 7. Orbiter FEM Components

RESULTS

A dynamic analysis was performed on the Orbiter and launch stack to check for low natural frequencies. The suggested launch frequency minimums from the Boeing design guide (ref. 3) of 10 Hz lateral and 27 Hz axial were used. The launch stack minimum modes, shown in figures 8 and 9, were 10.5 Hz lateral and 27.8 Hz axial respectively. Figure 10 is the first mode shape of the Orbiter aeroshell at 54.6 Hz. No dynamic magnification factors were considered during launch.

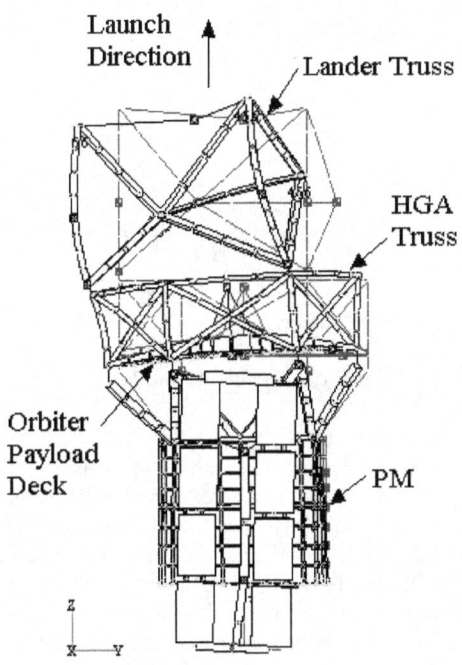

Figure 8. Lateral 10.5 Hz Launch Stack Mode

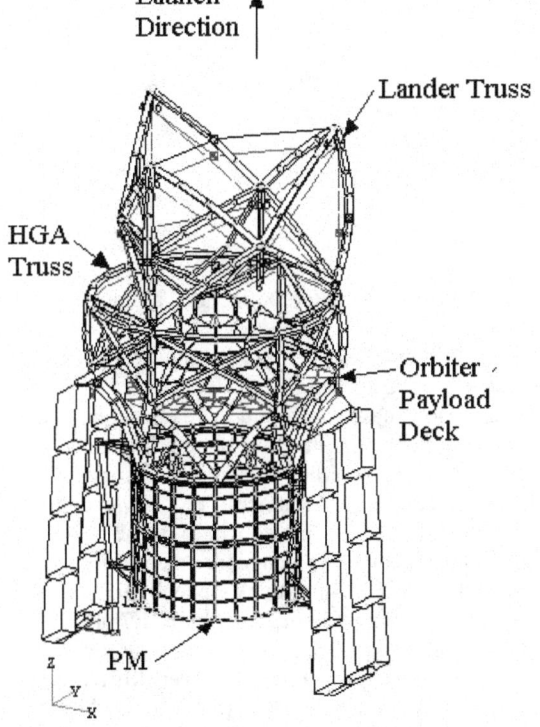

Figure 9. Axial 27.8 Hz Launch Stack Mode

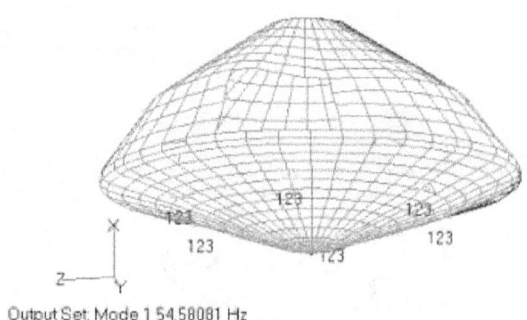

Output Set: Mode 1 54.58081 Hz

Figure 10. Orbiter 1st mode

The overall maximum deflections on the heatsheild were checked at launch and during entry. The largest deflections occurred on the heatshield during launch and were less than 1 mm. The exaggerated deflected shape of the heatshield is shown in figure 11. Double curvature exists where the six adapter truss points attach through the heatshield. The deflections were small and not considered a concern for TPS bonding.

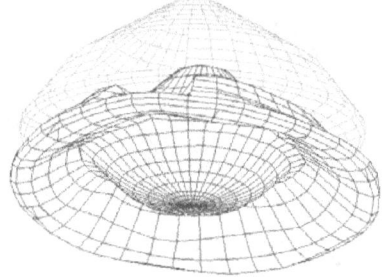

Figure 11. Orbiter Deflections at Launch

The Orbiter's aeroshell was sized after several iterations between nastran and Hypersizer™. The dominant load case was found to be during launch. The process optimized the aeroshell structure and indicated which materials and structural concept would produce the lightest aeroshell. Honeycomb core with face sheets were shown to be the best structural concept. The final core material for the heatshield was a 25.4mm thick Hexcell 5052 alloy hexagonal aluminum honeycomb with 1.7mm graphite polyimide face sheets. This design was similar to the Mars Exploration Rover (MER) heatshield except for the six hard points used in attaching the Orbiter heatshield to the PM. The backshell sizing done within HyperSizer™ showed a honeycomb core face sheet concept produced the minimum structural mass. The core was a 12.7mm thick Hexcell 5052 alloy hexagonal aluminum honeycomb and graphite polyimide face sheets of varying thickness. The cap plate design was similar to the backshell.

Results from analyzing the HGA and Lander trusses showed the optimal material was 2" OD, 0.12" wall M55J/954 tubes. The Orbiter/PM adapter truss was similarly made with 3.2" OD, 0.2" wall M55J/954 tubes. The sizes were driven by finding sections large enough to prevent buckling.

A summary of the final Orbiter aeroshell mass is given in table 1 and the total launch stack mass summary is shown in table 2. The total spacecraft mass for the launch configuration was 3173.2 kg and included the Lander, Orbiter and PM. TPS and non-structural masses were included plus allowances for miscellaneous items such as heatshield to backshell separation components. The design relies on a system of composite M55J trusses that form a load path into the PM. This system produced minimal displacements during launch and held stresses within safety limits that were: 1.4 on ultimate, 1.25 on yield limit and 1.5 for buckling.

Part	*Area*	Structure Mass	TPS Mass	*NSM*
Heatshield (TufRoc)	12.58 m²	41.58 kg	148.62 kg	0
Backshell (SLA-561V)	15.01m²	43.27 kg	28.69 kg	2.38 kg
Pallet Ring	1.20m²	42.47 kg	0	1.20 kg
Separation Ring	1.79m²	11.35 kg	0	.89 kg
Separation Ring Attachments	.45m²	2.85 kg	0	4.50 kg
Totals		141.52 kg	177.31kg	8.97 kg

Total Aeroshell (structure + TPS + NSM) = **327.80 kg**

Table 1. Orbiter Aeroshell Mass

Lander	400 kg
Lander Truss	61.8 kg
Orbiter Aeroshell + Payload	1200 kg
Orbiter/PM Truss	61.4 kg
PM	1450 kg
Total spacecraft	**3173.2 kg**

Table 2. Launch Stack Mass

CONCLUSION

The success of an aerocapture mission at Titan greatly depends on the mass reduction of the structure and the configuration of the launch stack. The design efforts encountered during this conceptual study showed the importance of defining the configuration in reducing spacecraft mass. The final launch configuration used an unconventional method of attaching the Orbiter heatshield to the PM. This method allowed a continuous load path from a 400kg Lander, into the Orbiter, through the Orbiter heatshield and into the PM. Maintaining a load path through the trusses that avoided the Orbiter aeroshell from supporting the Lander minimized the Orbiter aeroshell. The stacking arrangement also minimized the buckling lengths of the truss members as well as the number of required members. HyperSizer^{TM} was used to perform optimization sizing of the Orbiter aeroshell without a detailed mesh and extensive remodeling effort. The results indicated a honeycomb face sheet composite could produce a light structure while providing the necessary stiffness to meet minimum dynamic frequency requirements at launch. The results from this study have established a starting point for a detailed fea of the Orbiter aeroshell. Such an analysis could include varying core and lay-up thickness and detailed analysis of attachment connections and separation mechanisms. The structural mass for this design was within the mass margin estimated for a successful Titan mission.

ACKNOWLEDGEMENTS

The author recognizes the following for their contribution to the Titan structural analysis: Jonathan Lam (JPL) for his nastran modeling of the PM and Orbiter payload structure, Rob Bailey (JPL) for packaging the Orbiter and detailing payload requirements, Bernie Laub (ARC) for TPS sizing on the Orbiter heatshield and backshell, and
Eric Dyke (Swales Aerospace-LARC) for investigation of the Orbiter/Lander truss and stacking configurations.

REFERENCES

1. G. A. Hrinda, "Integrated Structural Analysis of the X-43 Using HyperSizer Sizing Software", CS/PSHS/APS Joint Meeting, November 2000.

2. G. A. Hrinda, "Structures For the 3rd Generation Reusable concept Vehicle", Space 2001 Conference

3. "Delta IV Payload Planners Guide", The Boeing Company, Huntington Beach, CA, 2000.

4. Collier Research Corporation, "HyperSizer Tutorial & Applications", 2nd edition, Collier Research Corporation, 1998.

5. C. S. Collier, P. W. Yarrington and M. Pickenheim, "Design Optimization Using HyperSizer", Collier Research Corporation, The 1998 MSC Americas Users 'Conference', Universal City, CA, Oct 5-8, 1998.

6. C. S. Collier, P. W. Yarrington and M. Pickenheim, "Next Generation Optimization Today", The 1997 MSC Aerospace Users Conference', Newport Beach, CA, November 18-20, 1997.

PLANETARY PROBE MASS ESTIMATION TOOL DEVELOPMENT AND ITS APPLICATION TO TITAN AEROCAPTURE

R. Eric Dyke
Swales Aerospace
Hampton, VA 23681

ABSTRACT

An integrated mass properties estimation and aeroshell structural sizing design tool, called PROBECODES, is being developed for quick sizing of conceptual, atmospheric-entry planetary exploration vehicles. This tool will eventually integrate EDS PLM Solutions I-deas® entry system geometry models, Microsoft® Excel-based entry system mass properties estimation spreadsheets, CFD generated aerothermodynamic loads, and structural sizing/optimizing software such as EDS PLM Solutions I-deas® or Collier Research's HyperSizer™ in order to create preliminary vehicle mass estimates, with particular emphasis on aeroshell structural sizing estimates. The Microsoft® Excel-based spreadsheet portion of the tool, called ProbeMAASS1 is currently being developed. It contains a mass properties/design database of previous planetary exploration vehicles and design studies; a mass properties breakdown and design description of these vehicles' major subsystems; mass estimation methods for the major subsystems of new vehicles using combinations of analytical techniques, empirical relationships, and user-specified data; and an automatic aeroshell structural finite element mesh generator. Currently this tool is limited to sphere-cone and ellipsled type vehicles, but future versions will include other types as they are developed. Though still under development, parts of this tool were successfully used to help quickly estimate Orbiter aerocapture system masses for a wide variety of TPS material/thickness and aeroshell size combinations for the Titan Aerocapture Systems Analysis study of 2002. In addition this tool was used to help estimate total launch wet mass for ten different Titan mission/launch vehicle configurations. This paper will present an overview of the full tool (PROBECODES) and an overview of the mass properties estimation spreadsheets (ProbeMAASS1). Lastly, it will present the mass sensitivity results from the Titan Aerocapture study.

NOMENCLATURE/ABBREVIATIONS

CBE	Current Best Estimate
CFD	Computational Fluid Dynamics
CG's	Centers of Gravity
EGA	Earth Gravity Assist
GDTL	Geometric Data Transfer Link
GUI	Graphical User Interface
L/D	Lift divided by Drag
LMDTL	Lumped Mass Data Transfer Link
LV	Launch Vehicle
MEL	Master Equipment List
MEM's	Mass Estimation Methods
MPD	Mass Properties Database
MPB	Mass Properties Breakdown
SEP	Solar Electric Propulsion
TPS	Thermal Protection System
VVVGA	Venus-Venus-Venus Gravity Assist
VGA	Venus Gravity Assist
ΔV	Velocity change

INTRODUCTION

As part of the ongoing exploration of the solar system, there is a need for an atmospheric-entry probe design tool that takes conceptual probe design data and provides quick-turnaround probe mass estimates and aeroshell structural sizing. To fulfill this need, an integrated mass properties estimation and aeroshell structural sizing design tool is currently being developed. This tool, called PROBECODES (Probe Conceptual Design Software), is comprised of several different components, including commercially available software, as shown in Figure 1. Current methods for probe mass estimation and aeroshell sizing often involve using separate, non-integrated historical data, mass estimation methods, and finite element modeling/structural sizing tools. Often the geometry models, mass estimations, and finite element modeling are done by different individuals using different tools which may or may not share data easily. The ultimate goal of PROBECODES is to integrate mission-driven payload requirements, EDS PLM Solutions I-deas[1] entry system geometry models, Microsoft Excel[2]-based entry system historical databases, Microsoft Excel-based mass properties estimation spreadsheets, CFD-generated aerothermodynamic loads (pressure and/or temperature distributions), structural materials data, and structural sizing/optimizing software such as MSC/NASTRAN[3], EDS PLM Solutions I-deas or Collier Research's HyperSizer[TM 4] into a seamless end-to-end tool in order to produce preliminary vehicle mass and aeroshell structural sizing estimates. The primary benefit of PROBECODES, when completed, over previous methods will be to allow a single user to quickly create conceptual probe geometry, estimate component masses, and size the aeroshell structure all within a single tool in a short time.

PROBECODES and the mass estimation component, ProbeMAASS1 (Probe Mass Approximations & Automeshing for Structural Sizing, version 1), are currently limited to sphere-cone and ellipsled shaped entry vehicles. Future entry vehicle shapes, such as bi-conics, bent bi-conics, and other configurations will be addressed in later versions. This tool also assumes a general payload packaging layout has been established. Lastly, the automated structural sizing aspects of the tool are currently limited to the aeroshell. Sizing for the wide variety of possible internal structural components such as payload support structure (decks, rings, etc.) is not currently supported. Such items must be handled in an "offline" manner, with the resulting finite element meshes and masses then being used as input to this tool.

While still in the development stages, portions of ProbeMAASS1 were successfully used for mass sensitivity studies for the Titan Aerocapture Systems Study of 2002[5]. Aeroshell mass sensitivities were developed for a range of provided TPS materials and aeroshell size/TPS material combinations. In addition, the tool was used to help estimate total (full stackup) launch wet masses for ten different mission architectures. The results of theses studies are presented later in this paper.

OVERVIEW OF PROBECODES TOOL

The baseline requirements in developing PROBECODES are that preliminary aeroshell structural design and probe mass estimate results be produced from the conceptual data in a relatively short time (2 to 3 days), and that these results be produced with a minimum of special hardware and software. The current tool architecture primarily uses EDS PLM Solutions I-deas, Microsoft Excel, and Collier Research's HyperSizer[TM]. I-deas is used for aeroshell solid modeling and surface area/volume calculations and for aeroshell structural finite element model editing and solving. Microsoft Excel is used for the probe mass properties database of historical probes and probe studies, the mass estimation methods, aeroshell surface area calculations, and aeroshell structural finite element model generation. Though not yet integrated into PROBECODES, Collier Research's HyperSizer[TM] will be used to optimize the aeroshell structure. As newer, better methods become available, or as the tool components allow themselves to be combined or separated, the tool architecture as shown in Figure 1 will be modified to incorporate those changes to produce a more efficient product.

PAYLOAD REQUIREMENTS, MISSION DESIGN

The user must start with a given or assumed mission profile. The mission profile determines the payload components required to meet the mission objectives, including scientific instrumentation and their associated support subsystems (telecom, navigation, command and data handling, power, etc.). The mission profile also determines the delivery time and atmospheric entry profile, including entry speed, entry angle, maximum deceleration, and aeroheating rates and total heat loads. The aerodynamic performance requirements, such as vehicle L/D, ballistic coefficient, etc., necessary to ensure the proper entry profile help determine the general class of entry vehicle. Currently, the low L/D vehicles are represented by the axisymmetric sphere-cone shapes, and the mid L/D vehicles are represented by the ellipsled shape. Together, the general vehicle shape required and the volume necessary to accommodate the proposed payload determine the overall aeroshell size.

SOLID MODELING (EDS PLM SOLUTIONS I-DEAS®)

EDS PLM Solutions I-deas is used as the entry vehicle solid modeler. Multiple axisymmetric sphere-cone solid models have been created within an I-deas model file, including a stand-alone forebody, stand-alone backshells (conic, bi-conic, multi-conic, hemispheric, and spherical cap), and combined forebody/backshell models for bi-conic and hemispheric cap backshells. These models are parametrically dimensioned such that the user may edit one or all of the cross section dimensions and the solid model is automatically updated. An axisymmetric ellipsled solid model has also been created with a single backshell separation configuration. Additional solid models will be created for more general (non-axisymmetric) ellipsled shapes with more general backshell separation geometries.

GEOMETRIC DATA TRANSFER LINK (GDTL)

The GDTL transfers the solid model geometry data from I-deas® to the ProbeMAASS1 spreadsheets. With the appropriate solid model opened, an I-deas® program file is used to list and save the model dimensions and surface areas. The program file is then transferred (if necessary) to the PC platform where ProbeMAASS1 resides. A macro within ProbeMAASS1 then reads the file and uploads the dimensions and surface areas for use in later calculations.

ENTRY SYSTEM DATABASE

The entry system database is a compilation of planetary exploration probe and probe study design data. It is in Excel spreadsheet form, with separate sheets for the sphere-cone and ellipsled vehicle types. These databases, described in more detail in a later section, are part of the ProbeMAASS1 spreadsheets, and contain probe geometry, entry environment definition, subsystem and component masses, and structural materials and construction methods. The data are used to develop mass estimation methods in ProbeMAASS1, and allow the user to make quick initial guesses for subsystem or component masses based on similarity to previous designs.

USER INPUT PARAMETERS

User input parameters are values entered into the ProbeMAASS1 spreadsheets. They include such wide-ranging details as probe destination, number of propellant tanks, or parachute diameter. They are input in a "linear" fashion on each spreadsheet from top to bottom, as each spreadsheet is filled out for each probe system. ProbeMAASS1 uses the historical database, user input, and mass estimation methods (MEM's) discussed later to calculate component masses. The user can also override any or all calculated values in the spreadsheets. Since the ProbeMAASS1 spreadsheets will calculate aeroshell surface areas that are used in aeroshell structure and TPS mass estimation, the user may also input probe dimensions separate from the imported I-deas® geometry. This allows the user to take quick looks at the impact dimension changes have on aeroshell mass.

ENTRY SYSTEMS MASS PROPERTIES TOOL (PROBEMAASS1)

ProbeMAASS1 is the heart of this mass estimation tool. It will be discussed in more detail in a later section.

MASS PROPERTIES OUTPUT

The mass properties output currently consists of three items: the lumped mass data, the aeroshell mesh, and the probe mass summary. The lumped mass data consists of point masses and their associated centers of gravity (CG's) for larger components. They are either calculated by the ProbeMAASS1 spreadsheets or entered directly by the user, and are added to the aeroshell mesh file. While the spreadsheets currently determine the point masses, the CG calculations have not yet been fully integrated into ProbeMAASS1. The aeroshell mesh is a structural finite element mesh created by internal macros in the ProbeMAASS1 spreadsheets and is described in more detail in a later section. The probe mass summary sheet lists current best estimate (CBE) masses for each probe component. Component masses are also added to give associated subsystem masses and total probe mass. The user may also specify separate uncertainty factors to be applied to each component CBE to determine the growth mass. In the first pass through ProbeMAASS1, before aeroshell structural sizing has been completed, the aeroshell mass is calculated parametrically from historical probe data. Due to the wide variety of probe configurations, this first aeroshell mass estimate is considered low fidelity and must be updated after the structural sizing has been completed on the aeroshell.

INERTIAL, AERO & THERMAL LOADS

These loads are determined by external analyses. Simple inertial loads (gravity and body accelerations) and uniform aerodynamic pressures can be applied to the mesh within the ProbeMAASS1 spreadsheets and macros. Thermal loads/temperature distributions and more complex aerodynamic pressure distributions must be applied within I-deas® or some other pre/post processor at this time.

MATERIAL DATA

The material data for finite element analysis is typically specified by the user within a finite element pre/post processor. Since the aeroshell structure is to be optimized using Collier Research's HyperSizer™ (see below), the current tool architecture assumes the aeroshell will be optimized using a HyperSizer™ material database.

FINITE ELEMENT MODELING/STRUCTURAL SIZING

The aeroshell mesh is imported into I-deas® or other appropriate pre-processor. There the mesh is edited/modified, lumped masses are added, and secondary or payload support structure is added as necessary. The model is then solved in I-deas® or MSC/NASTRAN. The mesh input and results files are then submitted to HyperSizer™ for structural optimization of the aeroshell.

HyperSizer™ is a windows-based, commercially available structural sizing/weight estimation/weight optimization software tool[4]. A plate and/or beam element finite element model and its results file are first read into HyperSizer™. A graphical interface allows the user specify a range of structural "families" (i.e., unstiffened panels, sandwich panels, uniaxially stiffened panels), a range of structural "concepts" (i.e. "C" or "I" stiffeners), and a range of materials for various regions of the finite element model. HyperSizer™ then steps through each combination of designs for each region from lightest to heaviest until a successful design is created. The major advantage of HyperSizer™ is that it performs numerous user specified closed-form failure checks on each region for each design (family/concept combination) using the initial finite element model. This eliminates the need to create multiple finite element models.

Once aeroshell optimization is complete, the aeroshell mass is entered into the probe summary spreadsheet. Future enhancements will include automatic updates of the probe mass summary spreadsheet after aeroshell optimization.

STRUCTURAL DESIGN AND MASS ESTIMATE (PRELIMINARY)

As part of the HyperSizer™ optimization routine, several structural concepts can be evaluated, including sandwich construction, blade stiffened panels, isogrids, etc. Thus, an aeroshell mass and structural design are produced as part of the HyperSizer™ optimization process.

OVERVIEW OF PROBEMAASS1

ProbeMAASS1 is the main component of PROBECODES. It is a series of Excel spreadsheets which act as the "clearing house" for all probe design data, and is the primary user interface for probe mass estimation. It contains an import interface for aeroshell solid model dimensions and surface areas created in I-deas®. It also contains Visual Basic macros for generating aeroshell structural finite element meshes for sphere-cone and ellipsled geometries. The approach in developing ProbeMAASS1 was to use a linear progression through the series of Excel spreadsheets with a minimum of user inputs in order estimate the entry probe mass. Early on in the tool development, the Excel spreadsheet format was chosen over various "GUI" formats so that all historical data, references, supporting notes, and equations could easily be viewed by the user. Figure 2 shows the major functional features of ProbeMAASS1. These features are discussed in more detail in the following sections.

MASS PROPERTIES DATABASE (MPD)

The Mass Properties Database (MPD) is a collection of historical probe (flight) and probe study (paper) data. It contains a limited amount of data for European, Soviet, and Japanese probes, but is primarily a collection of data from NASA-directed probes. There are separate database spreadsheets for the sphere-cone and ellipsled type entry vehicles. The probe data are grouped according to probe destination (i.e. Venus, Mars, etc.), and are subdivided into the major categories shown in Figure 3. The first four categories are probe overviews: Mission Overview (destination, launch date, launch vehicle), Total Mass Properties (mass, moments of inertia), Geometry (overall forebody and backshell dimensions), and Entry Environment (speed, angle, maximum deceleration, maximum heat rate and heat load). Next come the probes major systems: Forebody System, Aftbody System, Deceleration Systems, and Payload. Each system is further broken down into the subsystems shown in Figure 3, which are described in more detail in the next section. This general breakdown is followed for both sphere-cone and ellipsled type entry vehicles. There is considerably less data in the ellipsled database simply due to the limited amount of ellipsled flight and study data available.

As with all other spreadsheets in ProbeMAASS1, the database spreadsheets follow a color-coding convention. For the database sheets, cells containing flight data are colored turquoise, while cells containing paper study data are colored light turquoise. Some values, like probe ballistic coefficient, which are not found specifically in the literature, are calculated from other data, and the resulting cells are colored tan. Once the color conventions are understood, they help the user understand at a glance the type of data he or she is dealing with.

Every cell or every line of cells in the MPD contains a cell comment that lists a reference for the data shown. For cases where no data was found, the associated cell was left blank. In other cases, conflicting values were found, such as for a heat load or entry deceleration. In such cases, the most recent data value was entered into the cell, and other conflicting values, with references, were listed in the cell comment. Cell comments are also used to include more in-depth information beyond what is showable in the given cell, such as facesheet or honeycomb thicknesses for sandwich panel aeroshells.

MASS PROPERTIES BREAKDOWN (MPB)

The probes' major system masses are further refined by the Mass Properties Breakdown (MPB). The MPB is not a separate spreadsheet, but is embedded in and part of the MPD. The actual refinement is also shown in Figure 3. For the forebody and aftbody systems, the probe mass is broken down to thermal protection, structure, separation system, thermal control, and miscellaneous subsystems and their components. The deceleration systems are further broken down into the pilot, main, and terminal parachutes and terminal descent (airbag) subsystems. Even within these subsystems is more detailed information such as parachute deploy speed, deploy altitude, etc. The payload system is further broken down into structure, science instrument, power, guidance/navigation, telecom, command and data handling, thermal control, harness, propulsion, and miscellaneous subsystems. Again, within these subsystems is considerable detailed component mass such as propellant tank size and material, etc. Some of the more detailed information is shown in the cells, while other is contained in the cell comments.

MASS ESTIMATION METHODS (MEM'S)

The bulk of ProbeMAASS1 is devoted to the Mass Estimating Methods (MEM's). These are the individual spreadsheets, as shown schematically in Figure 4, which are used to estimate the component, subsystem, system, and total probe mass. A separate spreadsheet is devoted to each major system and its associated subsystems and components. The MEM's take several forms, depending on the system or component being estimated. In some cases, the user may simply refer to the MPD and use an historical component mass which is most similar to the current design, and which may be independent of probe size. An example would be aeroshell separation fittings. While the number of fittings may vary, the basic fitting design may be relatively constant, thus allowing the user to apply the historical mass to the current design for a quick mass estimate.

In other cases, simple parametric scaling is used to estimate a new component mass. For example, a new parachute mass may be estimated by scaling a closely similar historical parachute mass by the ratio of the diameter squared. Such scaling assumes a constant areal density (kg/m^2) between the historical design and the new design. Areal densities are also used to estimate masses for aeroshell structure and TPS. Using the aeroshell surface areas either from the geometry imported from the I-deas® solid model or calculated by the ProbeMAASS1 spreadsheets, in conjunction with areal densities from historical data, the aeroshell structure and TPS masses can be calculated. The historical areal densities are generally curve fits of previous probe design data. Figure 5 shows an example of a curve fit used to give a first-pass estimate of forebody aeroshell structural areal density for sphere-cone type probes.

The highest fidelity mass estimation uses closed form solutions. For this tool, closed form solutions are primarily used to calculate sphere-cone surface areas, and to estimate mass for the propulsion system and its components. Standard textbook equations [6] are used to estimate propellant mass, pressurant mass, and propellant and pressurant tank masses. While this closed form method may produce higher fidelity mass estimates, it requires more user input than the other two methods.

As with any tool, care must be used to ensure that correct and reasonable input is being used, and that the output is reasonable. While historical masses may be used for some components, technological advancements often reduce the size and mass of these components. Such mass reductions are not currently predicted in the MEM's. In addition, several of the MEM's in this tool use curve fitting from only a few historical data points to develop scaling parameters. Such curves are often based on a very limited number of data points. The MEM's are updated to incorporate new data as it becomes available. To that end, this tool has been designed with a great deal of flexibility to allow the user to override estimated masses at any level in the process (component, subsystem, full probe).

Validation of the MEM's is an ongoing process. Closed form solutions for sphere-cone surface areas and propulsion component masses have been checked by hand to verify their accuracy. The surface area calculations for ellipsled geometry using the internal macros have been compared with I-deas® calculated values and match to within 0.05%. Curve fits of historical data and parametric mass scaling laws are constantly being reviewed to ensure the most accurate mass estimation methods.

AEROSHELL AUTOMESHER

The last major feature of ProbeMAASS1 is the automatic aeroshell structural finite element mesh generator and its associated surface area calculators for the sphere-cone and ellipsled aeroshells. The sphere-cone probe dimensions and surface areas can be imported through the GDTL discussed earlier. In order to provide more flexibility and to quickly evaluate the impact dimensional changes have on probe mass, the user may also enter override dimensions for the forebody and aftbody in their respective spreadsheets. Figure 6 shows generic sphere-cone forebody and aftbody cross-sections and dimensions. The spreadsheets then calculate surface areas using standard equations [7]. The I-deas® or override dimensions are then carried into the Sphere-Cone Aeroshell Mesh Generator spreadsheet (See Figure 4.) where the user then enters values to control the mesh density. Once all mesh control variables are entered, the internal macro generates a plate element structural finite element mesh with dummy material properties. The user may specify whether the forebody only, backshell only, or combined forebody and backshell meshes are created. For the sphere-cones, there is only one separation configuration: the backshell separating from the forebody at a given horizontal plane. The sphere-cone macro supports forebodies with hemispherical, spherical cap, conic, bi-conic, or multi-conic backshells. Figure 7 shows a typical mesh for a sphere-cone aeroshell with forebody and bi-conic backshell. The resulting mesh file is output as an MSC/NASTRAN .dat bulk data file. This format allows for easy model import into a wide variety of finite element pre/post processors (FEMAP, PATRAN, I-deas®, etc.). After import into the pre-processor, the user may edit the model as necessary to incorporate lumped masses and/or secondary structure or modify material properties, then solve the model. The user then sends the model and solution files to HyperSizer™ for structural optimization. Currently, the link to HyperSizer™ is a manual transfer. Future enhancements will include automated links.

For ellipsled entry vehicles, dimensions and surface areas can also be brought in through the GDTL. At this time, the GDTL only supports axisymmetric ellipsled vehicles with horizontal aft and canted forward backshell/forebody separation configurations. The GDTL will be updated later to transfer geometry data for more general ellipsled configurations. The ProbeMAASS1 Ellipsled Surface Area Calculator spreadsheet currently allows the user to enter dimensions for more general ellipsled geometries. Figure 8 shows a generic ellipsled with associated dimensions. Ellipsleds may be non-axisymmetric (upper and lower halves with different ellipse semi-axis ratios), and may have one of eighteen different backshell separation configurations. Similar to the sphere-cones, the dimensions here are carried into the Ellipsled Aeroshell Mesh Generator spreadsheet where the user enters all the required mesh control values, and the internal macro creates the aeroshell plate element structural mesh. The mesh control values allow the user to specify creation of the ellipsled forebody only, backshell only, base only, or all three. Figure 9 shows a typical ellipsled plate element mesh generated by the automesher. The resulting mesh can then be imported into a pre/post processor for editing and solving, then sent to HyperSizer™ for structural optimization.

FUTURE ENHANCEMENTS

ProbeMAASS1 is continually being updated. As new historical data becomes available, it is added to the database. This new data is then used to update the MEM's where applicable. Currently, the automatic mesh generators only create plate elements for the aeroshell. Future enhancements include a lumped mass data transfer link (LMDTL). Within the spreadsheets, centers of gravity (CG's) will be estimated for various point masses such as fuel tanks, electronics boxes, parachutes, etc. The user will also be able to override these estimates by inputting the desired CG's. The point masses and their associated CG's will be added to the finite element model when it is generated. In addition, more automated links are planned to open and transfer geometric data from the I-deas® solid model files into the ProbeMAASS1 spreadsheets, and to access and run the structural finite element codes such as I-deas® and HyperSizer™.

APPLICATION TO TITAN AEROCAPTURE SYSTEMS STUDY (2002)

Though still under development, parts of this tool were successfully used in conjunction with other subsystem mass estimates[8] provided by the Jet Propulsion Laboratory (JPL) and TPS sizing[9,10,11,12] provided by Ames Research Center (ARC), Langley Research Center (LaRC), and Applied Research Associates (ARA) to help estimate Titan orbiter masses for a wide variety of TPS material/thickness and aeroshell size/TPS combinations and to help quickly estimate total wet launch masses for the Titan Aerocapture Systems Analysis study of 2002.

IMPACT OF TPS SELECTION

The goal of the TPS selection sensitivity study was to determine the impact of aeroheating levels and, hence, TPS material selection and thickness on the total Titan orbiter mass, the aerocapture system mass, and the orbiter forebody TPS mass. In addition, the impact on aerocapture system mass fraction and orbiter ballistic coefficient were also investigated. This study required separate detailed aeroheating and TPS analyses for a 6.5 km/s Titan atmosphere entry to determine aeroheating rates and loads (both radiative and convective), and TPS material selection and thickness. For this study, several assumptions were made:

1. The orbiter aeroshell was held at a constant 3.75 meter maximum diameter.
2. The backshell TPS material (SLA) and areal density (1.87 kg/m^2) were held constant[9].
3. The orbiter wet mass (without aerocapture system) [8] was held constant.
4. The aeroshell structural areal density (kg/m^2) [13] was held constant.
5. The forebody TPS thicknesses and areal densities, shown in Table 1 below, were held constant for each case. All values except C-C data are from References 9-11. C-C data is from Reference 12.

Material	Thermal Load Case	Thickness (cm)	Areal Density (g/cm^2)
SRAM-14	Nominal lift down	1.55	0.348
SRAM-17	Nominal lift down	1.93	0.526
SLA-561V	Nominal lift down	2.43	0.622
SRAM-20	Nominal lift down	2.08	0.667
SRAM-20	0.8 conv, 1.8 rad	2.54	0.814
SRAM-20	0.6 conv, 2.6 rad	3.76	1.204
PhenCarb-20	Nominal lift down	2.34	0.711
PhenCarb-20	0.8 conv, 1.8 rad	2.71	0.868
PhenCarb-20	0.6 conv, 2.6 rad	2.91	0.931
TUFROC	Nominal lift down	5.13	1.181
TUFROC	0.8 conv, 1.8 rad	5.70	1.289
TIFROC	0.6 conv, 2.6 rad	6.12	1.371
C-C (Genesis-style) +15%	Nominal lift down	5.51	1.493
C-C (Genesis-style) +15%	0.8 conv, 1.8 rad	5.99	1.587
C-C (Genesis-style) +15%	0.6 conv, 2.6 rad	6.39	1.665

Table 1. Titan Aerocapture TPS Sizing Data

The nominal orbiter dimensions were adjusted for each material and thickness in order to calculate surface areas at the mid-thickness of the forebody TPS. These dimensions were entered into the Sphere-Cone Forebody spreadsheet in ProbeMAASS1 (See Figure 4.) as user input, and the spreadsheet calculated the associated surface areas. TPS areal densities were also entered later as user input, and the TPS mass was calculated by the spreadsheet. Fifteen TPS material/thickness combinations were evaluated and TPS masses calculated in just a couple hours. The resulting masses were added to other previously calculated orbiter system masses to determine total orbiter wet launch mass, aerocapture system mass, aerocapture system mass fraction, forebody TPS mass, and aerocapture ballistic coefficient as a function of TPS. The above items were also calculated for TPS with 30% and 50% mass margin. The results were plotted as bar graphs[14] as shown in Figures 10 through 14. The TPS material TUFROC was chosen as the baseline. ProbeMAASS1 allowed the user to show the orbiter mass sensitivity for a wide variety of TPS materials and thicknesses in a very short time. For this particular study, the lightest candidate investigated (SRAM-14) showed a potential 128 kg mass savings over the baseline TUFROC TPS material.

IMPACT OF ORBITER DIAMETER/TPS COMBINATION

The goal of the orbiter diameter/TPS combination sensitivity study was to determine the impact of aeroshell diameter/forebody TPS combination on total Titan orbiter mass, aerocapture system mass and mass fraction, orbiter forebody TPS mass, and orbiter aerocapture ballistic coefficient. Several assumptions were made for this study:

1. The 3.75 meter diameter aeroshell was photographically scaled down by 5%, 10%, 15%, and 20%. Payload packaging was not re-evaluated.
2. The backshell TPS material (SLA) and areal density (1.87 kg/m^2)[9] were held constant.
3. All Titan orbit dry mass subsystem masses were held constant except structure, which was reduced linearly with diameter.
4. The aeroshell structural areal density (kg/m^2)[13] was held constant.
5. The orbiter separation ring and payload pallet ring structure masses[13] were scaled linearly with diameter.

Similar to the study of mass sensitivity to TPS material alone, the modified aeroshell dimensions and TPS areal densities were input into the ProbeMAASS1 spreadsheet which calculated TPS masses. The TPS masses and orbiter systems masses scaled per items 4 and 5 above were added to other previously calculated orbiter system masses to determine total orbiter wet launch mass, aerocapture system mass, aerocapture system mass fraction, forebody TPS mass, and aerocapture ballistic coefficient for the various diameter/TPS combinations. The results for the five aeroshell diameters and six TPS materials under nominal aeroheating loads were plotted as bar graphs[14] as shown in Figures 15 through 19. The 3.75 meter diameter aeroshell with TUFROC TPS material was chosen as the baseline. Again, ProbeMAASS1 allowed the user to show the orbiter mass sensitivity for a wide variety of aeroshell diameter/TPS combinations in a very short time. This particular study showed the potential mass savings associated with reducing aeroshell diameter and showed significant mass variations with respect to aeroshell diameter and TPS, with the 3.0 m aeroshell/SRAM-14 TPS combination having 19.2% of the TPS mass of the baseline 3.75 m/TUFROC combination.

IMPACT OF MISSION CONFIGURATION/LAUNCH VEHICLE ON TOTAL LAUNCH GROWTH WET MASS

The goal of this sensitivity study was to determine the impact mission configuration/launch vehicle had on total launch growth wet mass. Several assumptions were made for this study:

1. Two all-chemical missions and eight aerocapture mission architectures, with either chemical or SEP (Solar Electric Propulsion) cruise propulsion systems[15], were evaluated. See Table 2 below.
2. For the aerocapture missions, a constant 515 kg aerocapture system growth mass and a constant 875 kg payload growth mass (in Titan orbit) was assumed.
3. For the all-chemical missions, a constant 937 kg payload growth mass (in Titan orbit) was assumed.
4. The EGA mission with SEP and aerocapture and a Delta 4450 launch-vehicle was the baseline.

Mission	Launch Vehicle	Gravity Assist	Propulsion System	Cruise Time (yrs.)	V_{entry} (km/s)	$\Delta V_{trajectory}$ (km/s)
1	Delta 4450	VVVGA	Chemical	9.3	6.2	1.5
2	Delta 4450	VGA	SEP	5.7	6.5	8.3
3	Delta 4450	EGA	Chemical	7.5	5.3	1.6
4	Delta 4450	EGA	SEP	5.9	6.5	6.7
5	Delta IV H	VVVGA	Chemical	9.3	6.2	1.5
6	Delta IV H	VGA	SEP	5.7	6.5	6.9
7	Delta IV H	EGA	Chemical	7.1	6.6	0.5
8	Delta IV H	EGA	SEP	6.1	6.5	4.6
9	Delta IV H	VVVGA	All chemical	~11.3	6.2	4.8
10	Delta IV H	VVVGA	All chemical	~11.9	5.9	3.5

Table 2. Mission Profiles[15]

ProbeMAASS1 was used to a lesser extent for this sensitivity study, but still proved valuable for producing the desired mass estimations in a very short time. For this study, the payload mass was assumed a constant 1390 kg for the aerocapture missions, and a constant 937 kg for the all-chemical missions. The trajectory ΔV's were used in the ProbeMAASS1 Propulsion spreadsheet to estimate required propellant mass (fuel and oxidizer for the assumed bi-propellant, pressure regulated system) and propellant tank size. The pressurant mass and pressurant tank size were estimated in the spreadsheet from the propellant tank sizes and user input for pressurant molecular weight and system pressure. The actual propellant and pressurant tank masses were estimated by using the tank areal density (kg/m^2) for similar radius tanks[16]. Thruster masses were estimated from user input for number and size of thrusters, and a thruster mass trend curve[6]. Miscellaneous propulsion system fitting mass was estimated as a user-specified mass fraction of the other propulsion system mass. Propulsion module dry mass was estimated by a simple linear scaling of a baseline chemical or SEP propulsion module, as appropriate. The propulsion system and propulsion module dry masses were then combined with other previously determined stackup component masses. Figure 20[14], shows the resulting total launch growth wet mass vs. mission configuration for the ten different configurations listed above. ProbeMAASS1 allowed the user to show the mass-saving benefits of aerocapture missions to Titan for a variety of missions in a very short time. These particular results show that the aerocapture missions to Titan provide significant mass savings over all-chemical missions primarily due to the large propellant and propulsion module dry masses required for the all-propulsive ΔV maneuvers.

CONCLUDING REMARKS

The PROBECODES tool and its major component, ProbeMAASS1, are still in development. The goal is to have a near-seamless end-to-end tool for taking conceptual atmospheric entry probe designs, estimating system, subsystem, and component masses, and performing structural sizing optimizations on the probe aeroshells to produce a preliminary probe aeroshell structural design, all in a relatively short time. While not yet completed, great strides have been made towards completing this process, and portions of the tool have been successfully used to support mass sensitivity trade studies for the 2002 Titan Aerocapture Systems Analysis study. Enhancements will continue to be made as the tool develops.

ACKNOWLEDGEMENTS

The author wishes to thank Dr. Mary Kae Lockwood of the NASA LaRC Vehicle Analysis Branch for her guidance in developing the tool architecture and tool focus. The author would also like to thank K. Chauncey Wu of the NASA LaRC Vehicle Analysis Branch for his inputs during the initial stages of the tool development, and Glenn A. Hrinda, also of the NASA LaRC Vehicle Analysis Branch, for his inputs for future tool enhancements.

REFERENCES

1. EDS PLM Solutions I-deas® software versions 9 and higher, Electronic Data Systems Corporation, Plano, TX
2. Microsoft® Excel User's Guides 1 and 2, Microsoft Corporation, 1994.
3. MSC/NASTRAN Quick Reference Guide, The MacNeal-Schwendler Corporation, 1992.
4. Collier Research Corporation, HyperSizerTM Structural Sizing Software, Book 1: Tutorial

& Applications, Second Edition, Collier Research Corporation, October 1998.
5. Lockwood, M.K., et al, Titan Aeroshell Aerocapture Systems Analysis Review, Jet Propulsion Laboratory, Pasadena, CA, August 29-30, 2002.
6. Brown, C.D., Spacecraft Propulsion, AIAA Education Series, 1996.
7. Staff of Research and Education Association, Handbook of Mathematical, Scientific, and Engineering Formulas, Tables, Functions, Graphs, Transforms, Research and Education Association, 1997.
8. Bailey, R.W., *Spacecraft Design*, Titan Aeroshell Aerocapture Systems Analysis Review, Jet Propulsion Laboratory, Pasadena, CA, August 29-30, 2002.
9. Laub, B., *Thermal Protection (TPS)*, Titan Aeroshell Aerocapture Systems Analysis Review, Jet Propulsion Laboratory, Pasadena, CA, August 29-30, 2002.
10. Congdon, W.M. and Curry, D.M., *2007 Titan Mission Analysis of ARA Ablators to Define Required Thickness and Weights*, Analysis Results of July 11, 2002.
11. Congdon, W.M., "Titan Ablator Thickness", Analysis Results of August 23, 2002, e-mail from W.M. Congdon (Applied Research Associates) to M.K. Lockwood (NASA LaRC).
12. Dec, J., "Genesis TPS for Titan Aerocapture", Analysis Results of September 27, 2002, e-mail from J. Dec (NASA LaRC) to M.K. Lockwood (NASA LaRC).
13. Hrinda, G.A., *Aeroshell Structure*, Titan Aeroshell Aerocapture Systems Analysis Review, Jet Propulsion Laboratory, Pasadena, CA, August 29-30, 2002.
14. Dyke, R.E., *Mass Sensitivities*, Titan Aeroshell Aerocapture Systems Analysis Review, Jet Propulsion Laboratory, Pasadena, CA, August 29-30, 2002.
15. Noca, M.A., *Mission Analysis*, Titan Aeroshell Aerocapture Systems Analysis Review, Jet Propulsion Laboratory, Pasadena, CA, August 29-30, 2002.
16. Sweetser, T.H., et al, "Titan Lander Conservative Science 01-06", JPL Team-X Report, June 15, 2001.

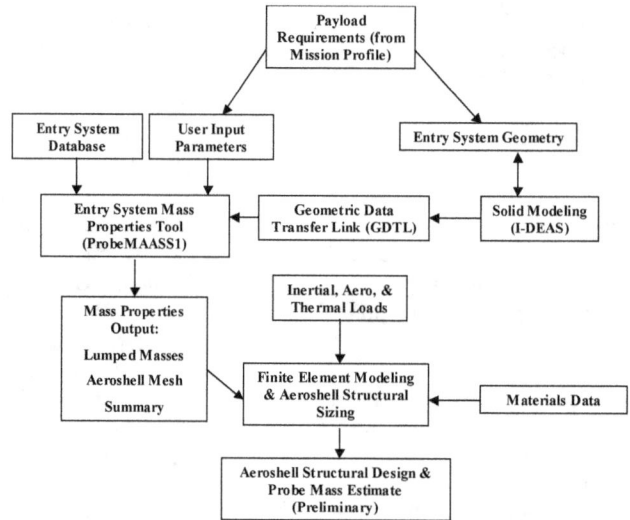

Figure 1. PROBECODES Entry System Mass Properties & Structural Design Tool Architecture

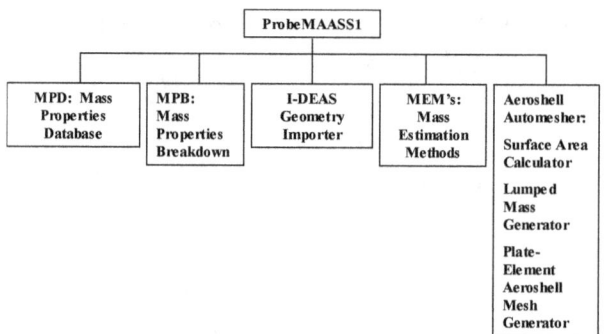

Figure 2. ProbeMAASS1 Functional Features

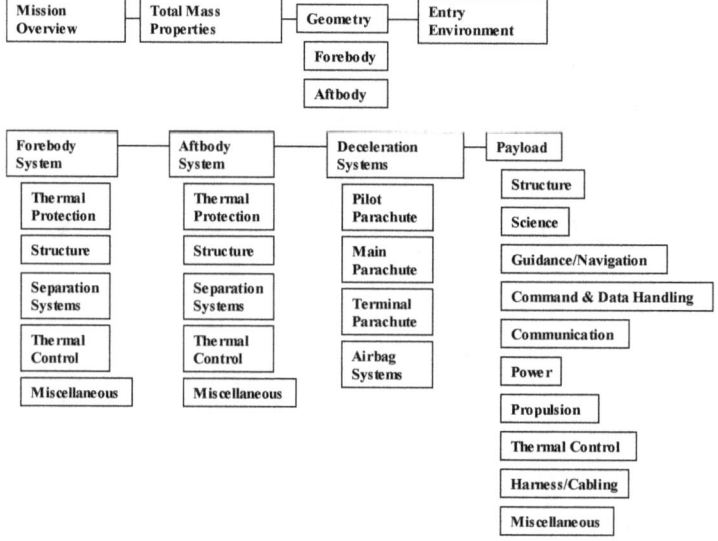

Figure 3. MPD/MPB Breakdown

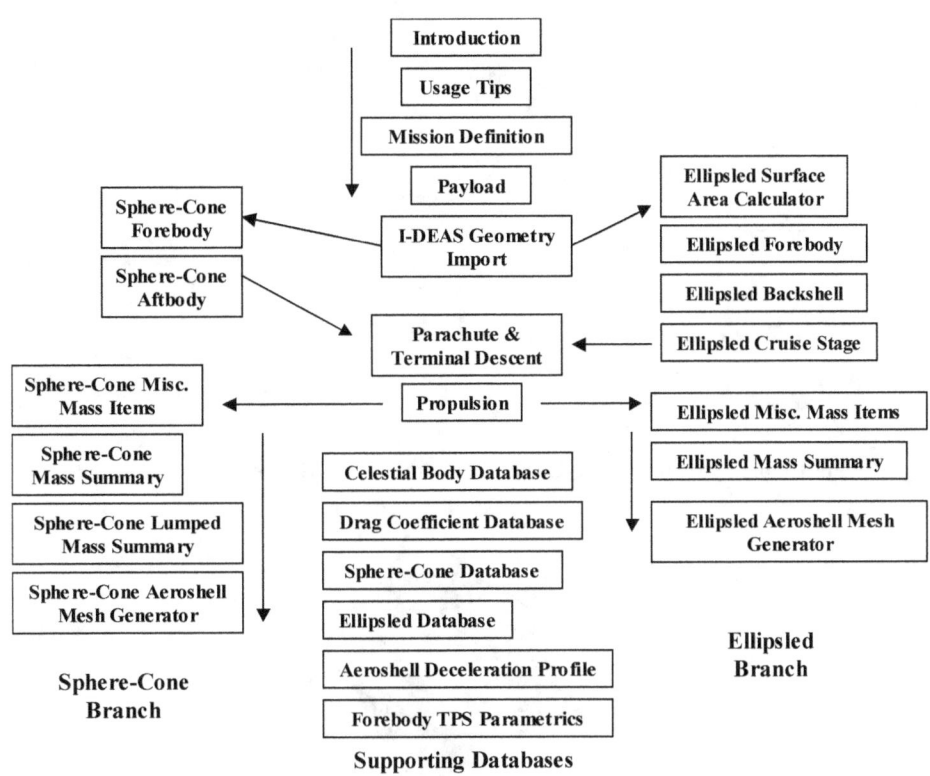

Figure 4. ProbeMAASS1 Component Spreadsheets

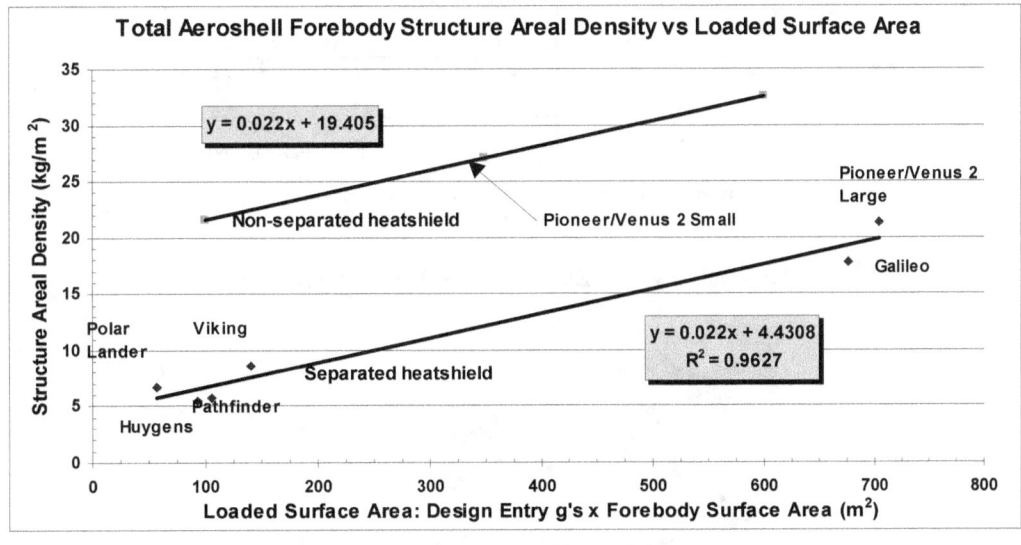

Figure 5. Sample MEM Curve from Historical Data

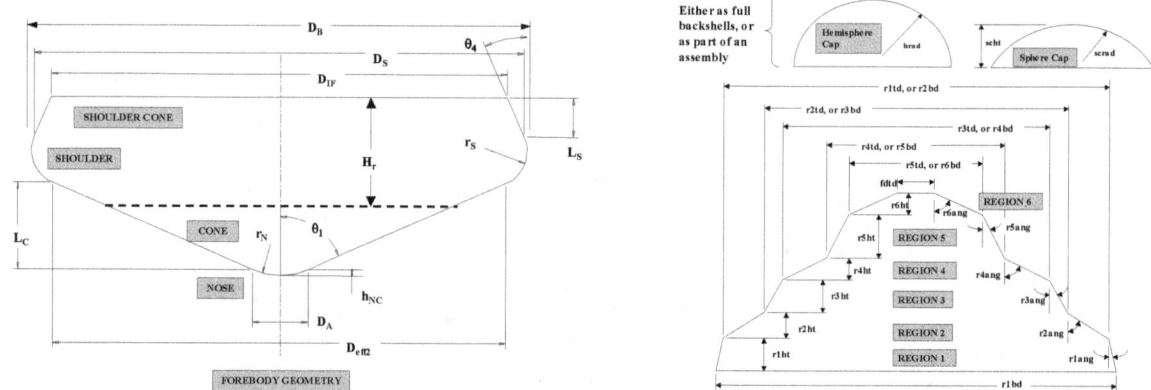

Figure 6. Typical Sphere-Cone Forebody and Aftbody Section Geometry

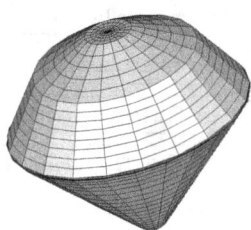

Figure 7. Typical Sphere-Cone Aeroshell Plate Element Mesh from Automesher

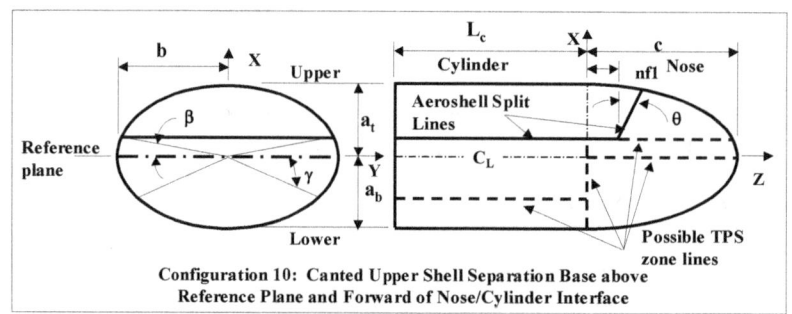

Figure 8. Typical Ellipsled Aeroshell Geometry

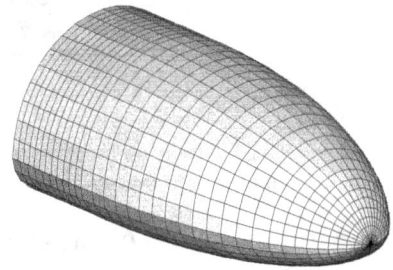

Figure 9. Typical Ellipsled Aeroshell Plate Element Mesh

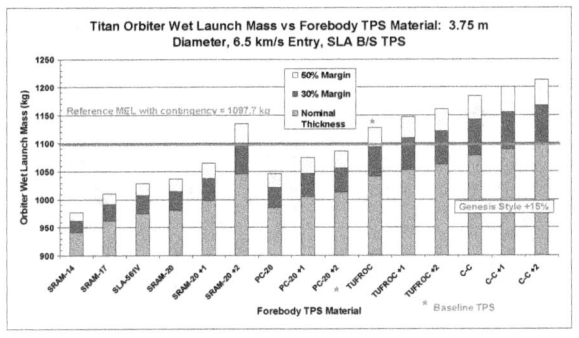

Figure 10. Orbiter Wet Launch Mass vs. TPS Material

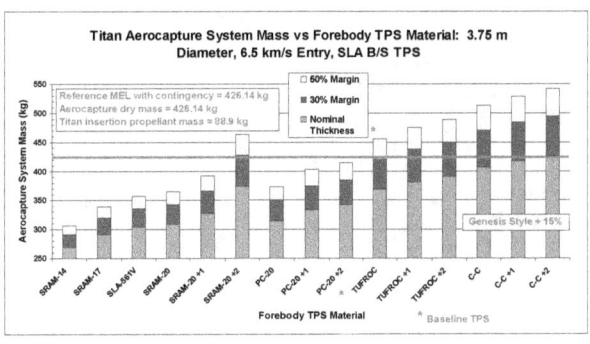

Figure 11. Aerocapture System Mass vs. TPS Material

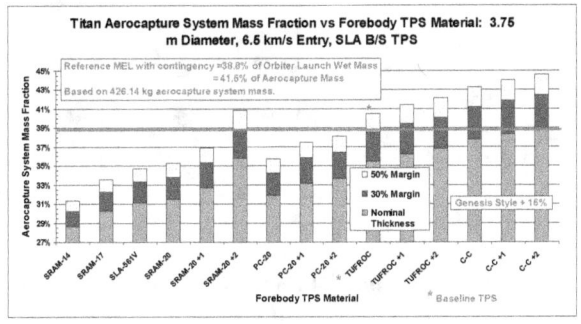

Figure 12. Aerocapture System Mass Fraction vs. TPS Material

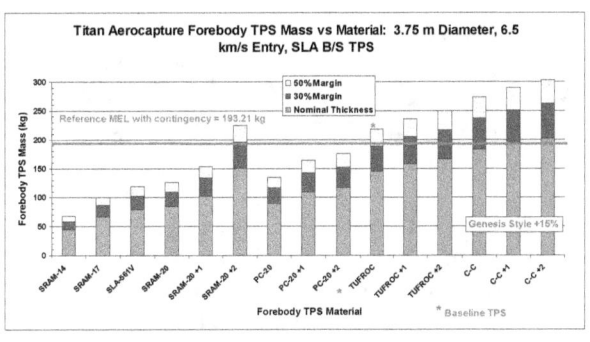

Figure 13. Forebody TPS Mass vs. TPS Material

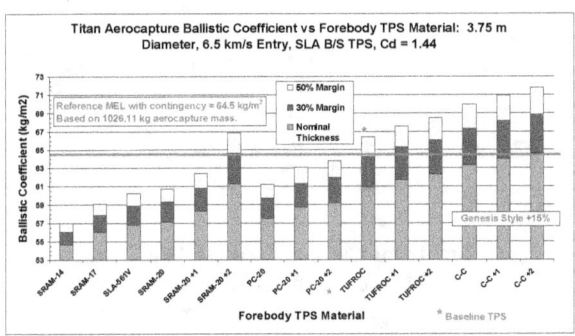

Figure 14. Aerocapture Ballistic Coefficient vs. TPS Material

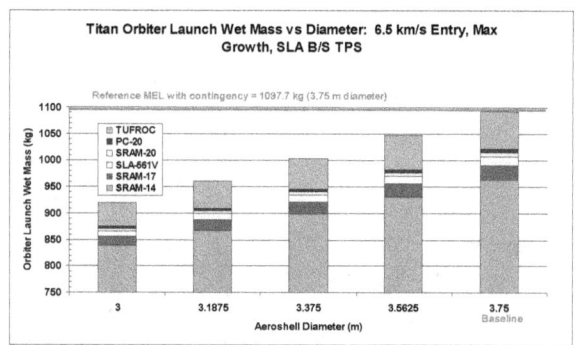

Figure 15. Orbiter Launch Wet Mass vs. Diameter and TPS

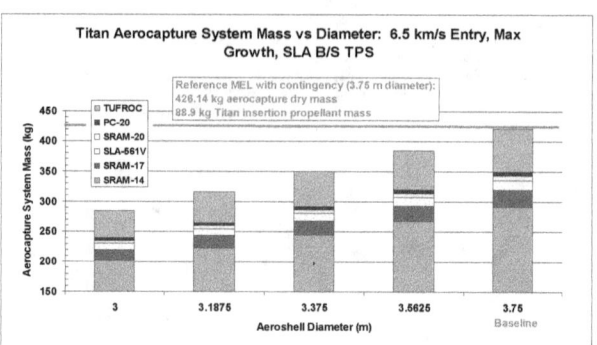

Figure 16. Aerocapture System Mass vs. Diameter and TPS

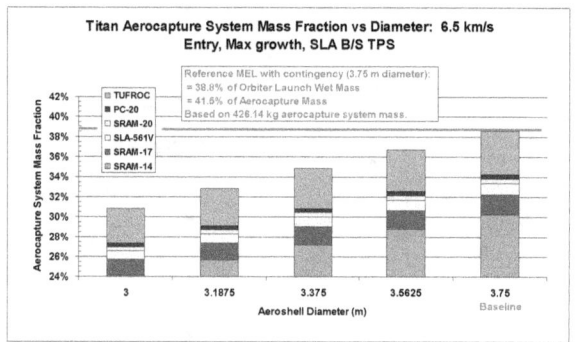

Figure 17. Aerocapture System Mass Fraction vs. Diameter and TPS

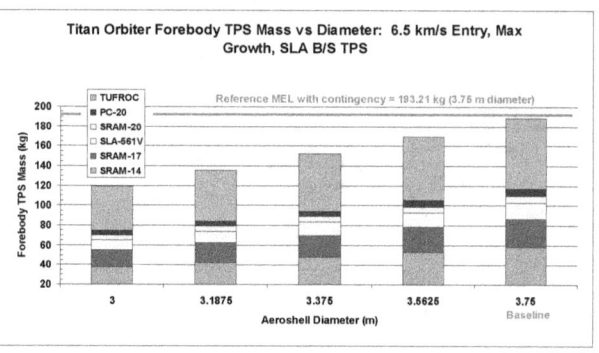

Figure 18. Forebody TPS Mass vs. Diameter and TPS

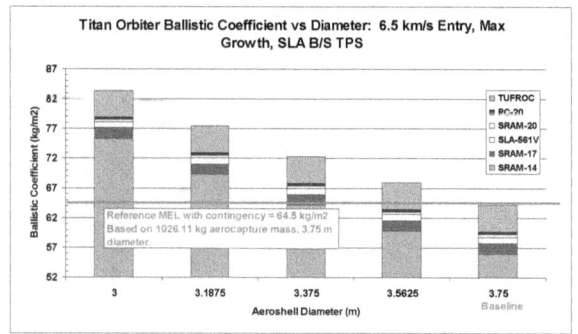

Figure 19. Orbiter Ballistic Coefficient Vs. Diameter and TPS

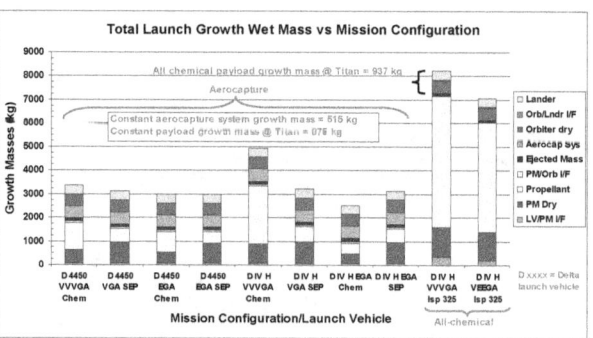

Figure 20. Total Launch Growth Wet Mass vs. Mission Configuration

www.ingramcontent.com/pod-product-compliance
Lightning Source LLC
Chambersburg PA
CBHW081727170526
45167CB00009B/3729